ROBO-ADVISORS AND ARTIFICIAL INTELLIGENCE IN HUMAN RESOURCES MANAGEMENT

Revolutionizing HR

ROBO-ADVISORS AND ARTIFICIAL INTELLIGENCE IN HUMAN RESOURCES MANAGEMENT

Revolutionizing HR

Sanjay Taneja, PhD
Swati Gupta, PhD
Girish Lakhera, PhD
Mohit Kukreti, PhD
Ercan Özen, PhD
Editors

AAP | APPLE ACADEMIC PRESS

First edition published 2026

Apple Academic Press Inc.
1265 Goldenrod Circle, NE,
Palm Bay, FL 32905 USA

760 Laurentian Drive, Unit 19,
Burlington, ON L7N 0A4, Canada

CRC Press
2385 NW Executive Center Drive,
Suite 320, Boca Raton FL 33431

4 Park Square, Milton Park,
Abingdon, Oxon, OX14 4RN, UK

Apple Academic Press exclusively co-publishes with CRC Press, an imprint of Taylor & Francis Group, LLC

Library and Archives Canada Cataloguing in Publication

CIP data on file with Canada Library and Archives

Library of Congress Cataloging-in-Publication Data

CIP data on file with US Library of Congress

ISBN: 978-1-77964-140-3 (hbk)
ISBN: 978-1-77964-141-0 (ebk)
DOI: 10.1201/9781779641410

About the Editors

Sanjay Taneja, PhD
Department of Management Studies, Graphic Era (Deemed to be University), Dehradun, Uttarakhand, India

Sanjay Taneja, PhD, is an Associate Professor (Research) at Graphic Era University in Dehradun, India, and an Affiliate Professor at Usak University in Turkey. With a focus on areas such as banking regulations, fintech, sustainable finance, and environmental finance, he holds a double master's degree in management with a specialization in green finance and marketing. He earned his PhD in Banking and Finance and completed his postdoctoral research on the impact of the European Green Deal on carbon emissions in Turkey.

He has published extensively in journals indexed by SCOPUS, Web of Science, SCI, ABDC, and UGC Care, with over 50 publications, including research papers, e-books, textbooks, and edited books. He has also published case studies in *Case Center*, ABDC-B, and Scopus journals, and holds several patents.

Dr. Taneja has a strong academic record, with numerous citations on Google Scholar and Scopus, and a history of active participation in conferences and faculty development programs (FDPs). He is frequently invited as a speaker and has served as a session chair, resource person, and judge at various conferences. He is known for his dynamic and positive approach, with a strong commitment to personal and professional growth.

Swati Gupta, PhD
Assistant Professor, PML SD Business School, Chandigarh, India

Swati Gupta, PhD, is an Assistant Professor at PML SD Business School, Chandigarh, India. She holds a doctoral degree in management and has over nine years of professional experience across various educational institutions. She specializes in teaching subjects such as consumer behavior, market research, marketing management, organizational behavior, sales and distribution management, services marketing, and retail management.

Having served as editor for various books under renowned publishing houses, including NOVA, Emerald, IGI Global, and AAP, she has contributed significantly to the academic field. Additionally, she has served as a reviewer for IGI Global and has authored Scopus-indexed chapters.

Girish Lakhera, PhD
Department of Management Studies, Graphic Era (Deemed to be University), Dehradun, Uttarakhand, India

Girish Lakhera, PhD, is a Professor in the Department of Management Studies at Graphic Era (Deemed to be University), India. With over 37 years of multidisciplinary experience spanning teaching, media, human resources, and administration, he brings a wealth of expertise to the field of editing. He has also served as Media In-Charge and Training and Placement Officer at Graphic Era University, contributing to academic research, editing, and publishing.

He is associated with IGI Global and Apple Academic Press as an editor. He has edited a book on deepfake technology for IGI Global, and another book is in process. His strong background in research, publishing, and media, coupled with years of leadership experience, positions him well for editorial work. Additionally, his role as a TV anchor and 20 years of service in the Indian Air Force have honed his skills in communication, management, and team collaboration.

Mohit Kukreti, PhD
Department of Business Studies, University of Technology and Applied Sciences (UTAS), College of Applied Sciences, Ibri, Oman

Mohit Kukreti, PhD, is presently associated with the recently established government University of Technology and Applied Sciences (College of Economics and Business Administration – Ibri). He served as Head of Department and Program Director for the IBA Program at the Directorate General of six CAS Colleges in the Ministry of Higher Education in the Sultanate of Oman. He holds a PhD in Human Resource Development (HRD) and has over 25 years of international experience in higher education administration, academic quality assurance, and strategic planning for higher education institutions (HEIs). He has held academic leadership positions in Nepal, Ethiopia, and the Sultanate of Oman.

Dr. Kukreti's academic contributions include numerous articles and chapters published in peer-reviewed, ABDC- and Scopus-indexed international Q1 journals and books. He has participated in and presented at conferences in various countries. He has conducted workshops, served as a session chair at conferences, undertaken editorial roles, and currently works as a reviewer for esteemed journal publishers such as Emerald, Routledge, IGI, and Inderscience. He has delivered talks at various universities, colleges, and to Ministry of Tourism officials, Government of India.

His expertise includes HEI quality assurance, HEI strategic management, and curriculum and course design at both undergraduate and postgraduate levels. He has served as a PhD examiner at several universities in India. His research areas span HRD, business management, HEI strategy development and quality assurance, entrepreneurship, and tourism.

Ercan Özen, PhD
Uşak University, Uşak, Turkey

Ercan Özen, PhD, is now a Professor of Finance in the Department of Finance and Banking, Faculty of Applied Sciences, University of Uşak, Türkiye (Turkey). He received his BSc in Public Finance (1994), MSc in Business-Accounting (1997), and PhD in Business Finance (2008) from the University of Afyon Kocatepe. His current research interests include various aspects of finance. He has served as co-editor for books published by eminent international publishing houses and has authored more than 100 publications. He has participated in many international conferences and is a board member of five international conferences and workshops. He is also the chair of the International Applied Social Sciences Congress and co-editor of two international journals—*Journal of Corporate Governance, Insurance, and Risk Management* (JCGIRM) and *Opportunities and Challenges in Sustainability* (OCS). In addition, he is a certified accountant, a member of the Aegean Finance Association, and a member of TEMA (Türkiye Erozyonla Mücadele, Ağaçlandırma ve Doğal Varlıkları Koruma Vakfı – Turkey Combating Soil Erosion, for Reforestation and the Protection of Natural Resources Foundation).

Contents

Contributors

Kapil Kumar Aggarwal
Chandigarh University, Gharuan, Mohali, Punjab, India

Rubel Amin
University School of Business, Chandigarh University, Gharuan, Mohali, Punjab, India

Nilesh Arora
University School of Business, Chandigarh University, Gharuan, Mohali, Punjab, India

Poonam Arora
University School of Business, Chandigarh University, Gharuan, Mohali, Punjab, India

Shwetank Avikal
Department of Management, Graphic Era Deemed to be University, Dehradun, Uttarakhand, India

Shaveta Bhatina
Center for Distance and Online Education, Manav Rachna International Institute of Research and Studies, Faridabad, Haryana, India

Nasima MH Carrim
University of Pretoria, Pretoria, Gauteng, South Africa

Sachin Ghai
Department of Management Studies, Graphic Era Deemed to be University, Dehradun, Uttarakhand, India

Tanuja Gour
School of Management, IMS Unison University, Dehradun, Uttarakhand, India

Medha Gupta
University School of Business, Chandigarh University, Gharuan, Mohali, Punjab, India

Munish Gupta
Chandigarh University, Gharuan, Mohali, Punjab, India

Swati Gupta
Assistant Professor, PML SD Business School, Chandigarh, India

Kshitiz Jangir
Manipal University Jaipur, Jaipur, Rajasthan, India

Dixit Kalra
Assistant Professor, School of Management Studies, CGC Mohali, Punjab, India

Ravneet Kaur
Chandigarh University, Gharuan, Mohali, Punjab, India

Uttam Kaur
University School of Business, Chandigarh University, Gharuan, Mohali, Punjab, India

Deepak Kaushal
Department of Management Studies, Graphic Era Deemed to be University, Dehradun, Uttarakhand, India

Rizwan Khan
University School of Business, Chandigarh University, Gharuan, Mohali, Punjab, India

Mohit Kukreti
Department of Business Studies, University of Technology and Applied Sciences (UTAS), College of Applied Sciences, Ibri, Oman

Pawan Kumar
Chandigarh University, Gharuan, Mohali, Punjab, India

Girish Lakhera
Department of Management Studies, Graphic Era (Deemed to be University), Dehradun, Uttarakhand, India

Navjot S. Miglani
University School of Business, Chandigarh University, Gharuan, Mohali, Punjab, India

Amrisha Minocha
Amity University, Noida, Uttar Pradesh, India

Amar Kumar Mishra
Department of Management Studies, Adamas University, Kolkata, West Bengal, India

K. K. Mishra
University School of Business, Chandigarh University, Gharuan, Mohali, Punjab, India

Larisa Mistrean
Department of Investments and Banking Activity, Faculty of Finance, Academy of Economic Studies of Moldova, Chisinau, Moldova

Nikita Suresh Nille
G. H. Raisoni College of Engineering and Management, Pune, Maharashtra, India

Nitin Pathak
University School of Business, Chandigarh University, Gharuan, Mohali, Punjab, India

Sakshi Raturi
School of Commerce, Graphic Era Hill University, Dehradun, Uttarakhand, India

Revati Ramrao Rautrao
Associate Professor, Dr D. Y. Patil B-School, Pune, Maharashtra, India

Atul Sharma
Chandigarh University, Gharuan, Mohali, Punjab, India

Ruchi Sharma
Department of Management Studies, Graphic Era Deemed to be University, Dehradun, Uttarakhand, India

Vikas Sharma
University School of Business, Chandigarh University, Mohali, India

Muhammad Rehan Shaukat
University of Technology and Applied Sciences, College of Applied Sciences, Ibri, Oman

Prashant Kumar Siddhey
University School of Business, Chandigarh University, Gharuan, Mohali, Punjab, India

Amar Singh
School of Commerce, Graphic Era Hill University, Dehradun, Uttarakhand, India

Archana Singh
Department of Management Studies, Graphic Era (Deemed to be University), Dehradun, Uttarakhand, India

Bhupinder Pal Singh Chahal
Yorkville University, British Columbia, Canada

Mandeep Singh
Chandigarh University, Gharuan, Mohali, Punjab, India

Muskan Singh
Department of Management Studies, Graphic Era Deemed to be University, Dehradun, Uttarakhand, India

Praveen Singh
Department of Management Studies, Graphic Era Deemed to be University, Dehradun, Uttarakhand, India

Sneha Singh
Department of Management Studies, Graphic Era Deemed to be University, Dehradun, Uttarakhand, India

Tejinder Singh
University School of Business, Chandigarh University, Gharuan, Mohali, Punjab, India

Anand Soni
School of Business, Bahrain Polytechnic, Isa Town, Bahrain

Sanjay Taneja
Department of Management Studies, Graphic Era Deemed to be University, Dehradun, Uttarakhand, India

Mohammad Taqi
School of Business, Bahrain Polytechnic, Isa Town, Bahrain

Jyoti Thakur
Center for Distance and Online Education, University School of Business, Chandigarh University, Gharuan, Mohali, Punjab, India

Sami Ullah
Faculty of Management Sciences, University of Central Punjab, Lahore, Punjab, Pakistan

Luan Vardari
University "Ukshin Hoti" Prizren, Prizren, Kosovo

Vivek Verma
Department of Management Studies, Graphic Era Deemed to be University, Dehradun, Uttarakhand, India

Abbreviations

ADF	augmented Dickey-Fuller
AI	artificial intelligence
APIs	application programming interfaces
AR	augmented reality
AT	apparent trust
ATS	applicant tracking system
AUM	assets under management
AVE	average variance extracted
BSM	Black-Scholes model
CA	Cronbach's alpha
CAGR	compound annual growth rate
COP	combined open interest predictor
CR	composite reliability
CVP	combined volume-based predictor
DL	deep learning
ERP	enterprise resource planning
ESG	environmental, social, and governance
EVP	employee value proposition
FAQs	frequently asked questions
FL	financial literacy
HR	human resources
HRI	human-robot interaction
HRIS	human resource information systems
HRM	human resource management
IoT	internet of things
IS	information systems
IT	information and technology
ITM	in-the-money
KMO	Kaiser-Meyer-Olkin
KPI	key performance indicator
ML	machine learning
NLP	natural language processing
NSE	National Stock Exchange
OI	open interest
OLS	ordinary least squares
OTM	out-of-the-money

PCR	put-call ratio
PDO	Petroleum Development Oman
PLS-SEM	partial least squares structural equation modeling
QC	quality control
R&S	recruiting and selection
RPA	robotic process automation
SEM	structural equation modeling
SHRM	Society of Human Resource Management
SLR	systematic literature review
SMBs	small and midsize businesses
TAM	technology acceptance model
TCS	Tata Consultancy Services
TPB	theory of planned behavior
TRA	theory of reasoned action
UF	user friendly
UTAUT	unified theory of acceptance and use of technology
VIF	variance inflation factor
VR	virtual reality

Foreword

As in any other sphere of organizational management, the integration of the principles and practices of human resource management (HRM) closely interacts with both technology and tradition. In this revolutionary age, *Robo-Advisors and Artificial Intelligence in Human Resources Management: Revolutionizing HR* can be seen as a guiding light for HR professionals and corporate executives, leading them toward a future that integrates innovation with human expertise.

During my work as a specialist in human resources and management, I have become familiar with the real impacts of robo-advisors on organizational productivity, staff satisfaction, and strategic planning. There is clear evidence that automation and artificial intelligence (AI) are fully capable of acting in parallel with—and in support of—the essential mission of defining the organizational environment and driving company performance by HR professionals.

Together, the academic conception of HRM as presented in *Revolutionizing HR,* the analysis of real-life case studies, and the implementation of best practices provide a vision of the future. There are numerous opportunities to enhance processes, increase productivity, and unlock people's potential in many aspects of operations—from recruitment to performance and talent management.

However, great creativity also requires great accountability. It is therefore important that we remain aware of the dangers, biases, and ethical implications of automation as we move toward a more advanced HRM environment. In *Robo-Advisors and Artificial Intelligence in Human Resources Management: Revolutionizing HR,* the authors provide guidance and encourage readers to think carefully before implementing ideas on the use of robo-advisors, ensuring that approaches remain democratic, transparent, and people-centric.

In the following chapters, readers will encounter a wealth of information, diverse perspectives, and practical tips that can help leverage the potential of robo-advisors to reinvent HRM processes and contribute to corporate success. This book provides much-needed guidance in the complex area where technology meets human capital in the 21st century—whether you are a practicing HR professional seeking ways to advance or a curious mind exploring the future of work.

Prepare yourself for an innovative, transformative, and enlightening journey. *Revolutionizing HR* opens the door to limitless opportunities in the field of HRM. It is now clear that the future of HRM lies at the heart of modern organizations.

—Raj Rajaram
Managerial Accounting and Financial Management,
School of Accounting, Economics and Finance,
University of KwaZulu-Natal, Pietermaritzburg,
KwaZulu-Natal, South Africa

Preface

Robo-Advisors and Artificial Intelligence in Human Resources Management: Revolutionizing HR offers a comprehensive analysis of the relationship between technology and the concept of human resource management (HRM) and their impact on the nature of work in the future. As editors of this pioneering work, it is our great honor to present it to readers.

In today's rapidly growing business environment, organizations are constantly searching for strategies that enhance efficiency, introduce innovative methods, and maximize the potential of their employees. This mission lies at the heart of HRM, which involves the systematic identification, development, and deployment of people within organizations to achieve maximum performance. However, with technology's continued expansion, traditional HRM strategies must now confront new opportunities and challenges.

This book is the product of extraordinary effort and the collaboration of experts, entrepreneurs, and scholars working within this complex domain. Centered on the concept of robo-advisors—automation supported by artificial intelligence—*Revolutionizing HR* provides an all-encompassing view of how the HRM environment is evolving, as illustrated through diverse perspectives, insights, and case studies.

Robotic advisors designed to enhance decision-making, simplify business processes, and increase organizational flexibility are becoming increasingly popular, transforming all subfields of human resource management (HRM), including selection and hiring, training, and development. However, their use also raises a series of questions, concerns, and ethical issues. What measures should be taken to ensure that these AI tools and algorithms are unbiased and free from prejudiced inclinations? What impact will they have on data protection, security, and privacy? What steps can be taken to overcome such challenges from the perspective of an HR specialist in today's dynamically evolving technological environment?

In the following pages, readers will discover valuable techniques, strategies, and recommendations for addressing these concerns and for expanding the potential of robo-advisors in the context of HRM. Whether you are an experienced HR manager interested in current trends in your field or an enthusiast eager to learn more about the future of work, this book will offer insight and inspiration.

I extend my deepest thanks to all the authors who have so generously shared their expertise and contributed their knowledge to this book. I am grateful for their extraordinary enthusiasm, hard work, and commitment to quality, which have elevated this project to a higher level.

I encourage you to read this book with an open mind, embarking on a journey of discovery, creation, and change. If you use *Revolutionizing HR* as a guide, its value is limitless as the future of HRM unfolds before our eyes. Thank you for joining me on this inquisitive journey into the future of work. Let the revolution begin.

—Editors

Introduction

Human resource management (HRM) can be considered a keystone of any enterprise or organization. In the rapidly evolving landscape of contemporary industry, HRM links an organization's strategies to the resources that form its backbone—its people. Thus, in observing modern knowledge in action and how it transforms business, HRM is no exception. It has now progressed to the phase of robo-advisors, marking one of the first instances of integrating artificial intelligence (AI) and mechanization into HRM practices.

The book titled *Robo-Advisors and Artificial Intelligence in Human Resources Management: Revolutionizing HR* explores this shift, analyzing the extent to which human skill and robotic capability can work hand in hand. It takes the reader on a journey from HRM's traditional foundations to its futuristic application of robo-advisors.

As passionate creators in the fields of HRM and modern technologies, we aim to explain how robo-advisors can be beneficial to HRM processes and what challenges and ethical issues are associated with this concept. Through a systematic study of the subject area, case studies, and expert insights, we seek to assist readers—especially human resource practitioners, managers, and anyone interested in returning HR to its basics—in undertaking this transformative endeavor with confidence and precision.

This volume is not merely a conceptual analysis; it is a practical reference for those prepared to understand the necessity and potential of change, incorporate technology, and reposition the HRM function for tomorrow's organizations. Whether you are actively engaged in the strategic management of human capital as the head of an HR division or an interested observer who occasionally wonders, "What is work?," *Revolutionizing HR* provides a framework for building a bridge between robo-advisors and organizational effectiveness—a comprehensive plan for creating a successful workforce in the 21st century.

—Sanjay Taneja, PhD
Swati Gupta, PhD
Girish Lakhera, PhD

CHAPTER 1

Transformative Talent Acquisition: Unleashing the Power of Robo-Advisors in HR for Smart and Efficient Hiring Processes

RAVNEET KAUR,[1] KAPIL KUMAR AGGARWAL,[1] ATUL SHARMA,[1] and SANJAY TANEJA[2,*]

[1]*Chandigarh University, Gharuan, Mohali, Punjab, India*

[2]*Department of Management Studies, Graphic Era Deemed to be University, Dehradun, Uttarakhand, India*

*Corresponding author

ABSTRACT

The advent of Artificial Intelligence (AI) and Robotic Process Automation (RPA) has profoundly recalibrated the paradigms of Human Resource Management (HRM), particularly within the domain of talent acquisition. Conventional recruitment frameworks, historically encumbered by protracted timelines, elevated transaction costs, and restricted geographical reach, are increasingly supplanted by algorithmic intelligence and robo-advisors that orchestrate candidate sourcing, pre-screening, and engagement with heightened precision. This chapter interrogates the transformative capacity of AI-embedded systems and RPA-enabled workflows, elucidating how eminent corporations such as IBM, Unilever, Amazon, and L'Oréal strategically deploy such technologies to expedite selection processes, cultivate inclusivity, and augment organizational agility. Anchored in an extensive literature review and conceptual synthesis, the analysis delineates both the salutary outcomes—efficiency gains,

Robo-Advisors and Artificial Intelligence in Human Resources Management: Revolutionizing HR.
Sanjay Taneja, Swati Gupta, Girish Lakhera, Mohit Kukreti, and Ercan Özen (Eds.)

DOI: 10.1201/9781779641410-1

predictive accuracy, and reduction in attrition—as well as the latent impediments, including algorithmic bias, erosion of privacy, and the psychosocial disquiet induced by potential job displacement. The findings underscore that while the techno-centric trajectory of HRM is fraught with ethical and structural dilemmas, its trajectory is inexorably converging towards augmented automation. Thus, AI and RPA emerge not merely as auxiliary tools but as catalytic agents reshaping the epistemology of recruitment and redefining the competitive calculus of contemporary enterprises.

1.1 INTRODUCTION

HRM is a domain that concentrates on examining human psychology, ethical standards, and human behavior. Every organization comprises a dedicated HRM department that handles and manages all employee-related factors such as recruitment, selection, performance appraisal, so forth. Although in recent years, the advancement of technology, such as Robo-Advisors and Robotic Automation, has significantly transformed the HRM landscape. Robo-Advisors in HRM use AI and robotics automation in cost- and time-effective recruitment and selection. This chapter focuses on understanding the AI and Robo upgraded process of recruitment, its benefits, and challenges faced by companies by quoting examples of companies using technologically upgraded techniques of hiring. This chapter also tries to draw a conclusion about the positive and negative impacts of the use of Robo-advisors in the hiring process. This chapter attempts to highlight the future of AI and Robo-advisors in personnel management.

1.2 REVIEW OF LITERATURE

1.2.1 PERSONNEL MANAGEMENT

Efficient personnel management empowers managers to establish strategies and recruit competent staff, while also aiding individuals in securing fulfilling employment with prospects for professional growth (Komakech, Obici, and Mwesigwa, 2021). Human resource management (HRM) consists of various factors and sub-factors such as hiring, maintaining, and training employees (Wall and Wood, 2005). HRM is crucial for any company's success and growth. Efficient and effective HRM practices save time and cost for organizations and help in bagging trained professionals (Aggarwal and Singhal, 2024). Hence, the hiring process plays a significant role in procuring and maintaining such talented employees in the present era. Given the significance of personnel management, the hiring process is at the core of HR practices, making it one of the most important activities in the success of organizations (Kok and Uhlaner, 2001). The recruitment process, once characterized by its lengthy duration and the heavy paperwork required for recruiters, has steadily undergone a revolution. The use of online recruitment has become widespread, indicating a transition towards a more efficient and streamlined method (O'Donovan, 2019). Therefore, personnel management can be defined as the process of managing and controlling human resources (HR) of organizations, such that it can bring efficient results for the organization without compromising personal growth. Current studies are progressively regarding HRM as a competitive advantage for firms, acknowledging employees as the central assets. The emphasis lies in comprehending the crucial significance of obtaining and overseeing these resources in achieving organizational triumph (Bas, 2012) (Figure 1.1).

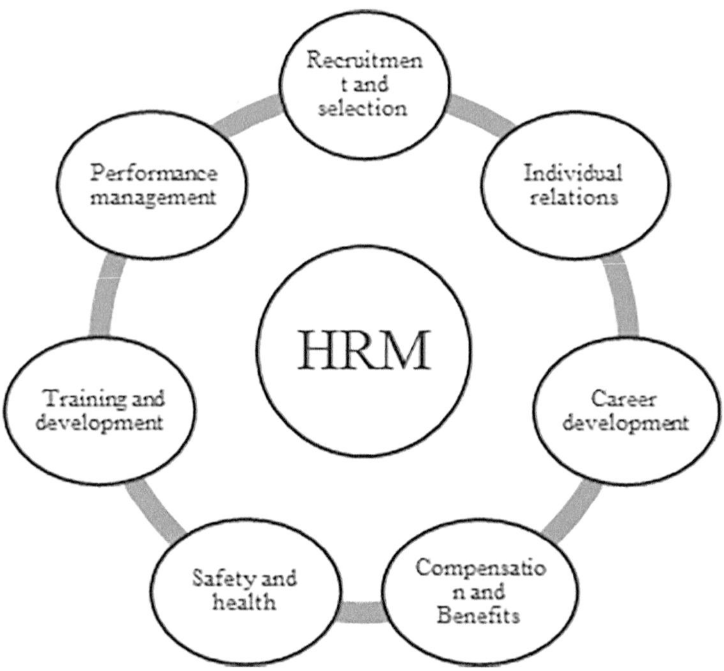

FIGURE 1.1 HRM's wheel.

Source: Author's Observation.

1.2.2 HRM RECRUITMENT ESSENTIALS

Hiring new talents is the process of encouraging competent employees in applying for jobs in the firm (Biswas, 2012). Griepentrog, Harold, Holtz, Klimoski, and Marsh (2012) advocated that recruitment is considered the central aspect of HRM, as it is the process of hiring people who would subsequently be subjected to other HRM activities. In order to progress, develop, and innovate, a firm has to hire competent employees with upgraded skills. The recruitment process ensures the generation of a pool of applications, so that the right person can be hired for the right job, with a primary emphasis on identifying and selecting the most qualified candidate for a particular post (Andersson et al., 2016). The sole purpose of recruitment is to actively search for and assess potential employees in order to secure their commitment and fill necessary positions for the effective functioning of the business (Aggarwal and Taneja, 2024).

1.2.3 CONVENTIONAL RECRUITMENT PROCESS

The conventional process of hiring lacks a standardized framework for its execution, instead being reported and theorized with subtle variations by numerous researchers (Acikgoz, 2019). Conventional recruiting, commonly known as "one on one" and theoretical-based recruitment, mostly relied on conventional media such as newspapers and manual job boards. Additionally, recruiters would meet potential applicants in specific venues to attract them (Chapman and Gødøllei, 2017). The traditional recruitment process has some common steps followed for the hiring process, but the process is very time-consuming. Lee (2005) stated that conventional hiring methods may be insufficient due to their lengthy hiring cycle, expensive cost per hire, limited geographical reach, and ineffective management of candidate relationships (p. 58). As mentioned above, there is no standardized framework or model for traditional recruitment, but even though different models have different stages, somehow, all models indicate the same levels. Many researchers added and provided different models for traditional recruitment. For this work, the authors decided to utilize the conventional recruiting model developed by Holm (2012). This model focuses on recruitment activities and provides detailed levels of information that can be later integrated with AI (Holm, 2012, p. 244).

Holm (2012, p. 244) proclaimed that the initial step in the hiring process is to conduct a job analysis to evaluate the requirements of the firm, and then generate a pool of applicants according to that. Secondly, to set up and select channels of recruitment such as newspapers, advertisements, so forth. Thirdly, pre-screen the applications received after registering them. Finally, communicate the results to the shortlisted applicants and move them for further assessments.

1.2.4 DIGITALIZATION OF HRM

This new era of technological advancements has transformed the way of doing business. The use of new hardware, software technologies, and smart gadgets would greatly enhance the efficiency of HR in performing their responsibilities, as well as significantly improve the overall quality of public everyday life. The advancements witnessed in technology over the period of time have significantly influenced all the stakeholders, employees, as well as HR managers (Pary et al., 2014). Hence, technology can function as a tool to aid and enhance organizations' decision-making process (Aggarwal and Agrawal, 2024), hence facilitating the transformation of organizational practices (Jatoba et al., 2019, p. 97). The advent of multinational culture, collaborations, and cutting-edge technologies such as Big Data, IoT, Machine Learning (ML), AI, and related advancements has significantly transformed the human lifestyle. Additionally, the emergence of Robotic Process Automation (RPA) applications has started to influence HR processes (Madakam, Holmukhe, and Jaiswal, 2019). According to Dhamija (2012), online hiring is widely regarded as one of the most popular alternative methods of recruiting, aimed at identifying and attracting highly skilled job candidates. This strategy benefits firms by reducing the cost, time, and effort involved in finding a skilled candidate for the job. Therefore, it can be said that technical advancements have revolutionized the HRM practices. Amongst these new advancements, the most evidently used are AI, Robo-advisors, and Chat Bots in recruiting and hiring processes of HRM.

1.2.5 CONCEPTUALIZING AI AND ROBO-ADVISORS

Understanding the concepts, development, and use of AI and Robo-advisors is crucial for this chapter, as this particular technological advancement has transformed the HRM practices of hiring and recruitment. AI is typically characterized as algorithms specifically created to mimic human behavior and cognition (Chenik, 2018, p. 1). John McCarthy (1998) defined AI as the discipline that involves creating intelligent devices, particularly intelligent computer programs. Prior scientific research has demonstrated that human intelligence not only contributes to the advancement of technology on our planet, but also acknowledges the intelligence of machines, thereby enhancing human intellect Salomon et al. (1991, p. 8). Nevertheless, sophisticated computer implementations, encompassing mathematics, logic, philosophy, languages, and decision theory, exhibit significantly greater complexity (Maini and Sabri, 2017, p. 9).

As per the definition provided by Dictionary 1, Artificial Intelligence (AI) refers to the magnitude of a mainframe or any other machine to conduct activities that are generally associated with human intelligence. Russell and Norvig (2009, p. 1) asserted that the progress in "AI" technology is a highly significant domain in the area of engineering, science, so forth, as it has the ability to do human duties with greater efficiency.

In recent times, there has been a growing popularity in the fields of data processing, ML, and AI. These advancements have significantly transformed the landscape of data and computing (Maini and Sabri, 2017, p. 3). These technical advancements are rapidly replacing humans in various roles and sectors. An appropriate example can be that Google developed an AI bot that has been trained to respond to inquiries, engage in discussions on ethics, and articulate its own viewpoint (Maini and Sabri, 2017, p. 3). One instance of AI implementation in the healthcare field is the rapid analysis of abnormal results in cancer patients' medical diagnoses (Maini and Sabri, 2017, p. 3).

AI in broad terms, refers to the capacity of machines to mimic human intelligence. It can be characterized by two key features: the capability to replicate human intelligence and the boundless capacity for learning (Raub, 2018, p. 531). Additionally, learning enables it to utilize virtual help in various functionalities.

Hence, AI is a comprehensive word that incorporates a diverse array of technologies. ML is the most commonly employed AI technology in business because of its extensive versatility, allowing it to be applied to an extensive variety of areas and also enhancing the value chain of processes. ML is the predominant technology utilized by 80% of Swedish organizations, while natural language processing (NLP) is used by 65%. Swedish firms have a larger representation in both technologies compared to the European average (Microsoft, 2018, p. 18).

1.2.6 AI AND TALENT ACQUISITION

The wave of technological advancements has surely eased the talent acquisition process. The expanding global economy has created an urge in organizations to hire talented and competent employees to lead. As stated in the Deloitte Global Human Capital Trends survey (2019, p. 71), due to the historically low unemployment rate and the abundance of skilled workers in several technological sectors, companies are now resorting to innovative branding strategies and boosting their utilization of AI to recruit candidates. Consequently, the task of locating and retaining highly qualified individuals has become progressively challenging.

The utilization of AI software in recruitment has witnessed a significant rise, resulting in substantial enhancements in the selection process of competent employees from an extensively generated pool of applicants (Sekhri and Cheema, 2019, p. 3074). An efficient AI-driven resume scrutinizing technology, known as an applicant tracking system (ATS), accurately identifies and assigns the most appropriate candidate to the relevant job position on the initial attempt (Kaur and Aggarwal, 2024). In addition, recruitment applications are currently utilizing chatbots driven by AI. Additionally, AI software proves beneficial in other ways, such as guaranteeing the usage of accurate vocabulary in job descriptions without making any presumptions about the gender of candidates, and precisely addressing the specific work needs (Rab-Kettler and Lehnervp, 2019, p. 107).

(Rab-Kettler and Lehnervp, 2019, p. 107) stated in their report that a significant portion of the hiring process is now carried out using automated techniques that do not require human participation. These include tasks such as creating job descriptions, screening CVs, arranging interviews, conducting initial screenings, making job offers, and facilitating pre-onboarding procedures. Nevertheless, the majority of organizations are using the following steps in the recruitment process by utilizing AI. There are various advantages of using AI-enabled hiring processes and software (Figure 1.2).

FIGURE 1.2 Recruitment process.

Source: Author's Observation.

Another example is that of L'Oreal company. L'Oreal implemented Chabot, Mya, which was used to screen applicants. Mya, as a chatbot and language processor, interacts with job applicants by asking questions and responding to their queries about the company's environment and cultural values. After Mya finishes screening, it shortlists the most and least suitable candidate list (Sharma, 2018). Therefore, it can be advocated that the inculcation of AI in HRM hiring practices has actually benefited the company, and the adoption of these techniques in the hiring process is rapidly increasing.

1.2.7 ROBOTIC PROCESS AUTOMATION (RPA) FOR RECRUITMENT

In the shifting landscape of HRM, the infusion of technology has become a game-changer, and one notable advancement is the integration of RPA in recruitment processes. RPA, a technological innovation that automates operations based on predefined rules, is currently revolutionizing the way firms find, attract, and recruit skilled individuals. Robotic automation is a software technology that hires robots to substitute HR operations in order to carry out administrative chores. RPA software smoothly works in combination with current tools to automate the activities, hence creating time- and cost-effective processes, in addition to efficiently completing specific jobs.

Businesses are incorporating RPA technology into their operations to streamline usually routine procedures and reduce associated costs. However, this technological process has both positive and negative impacts on the attitude of employees. Robots are electromechanical devices programmed by a computer to carry out a series of predetermined tasks. Automation is a systematic method of achieving full automation in a process. RPA, or Robotic Process Automation, is advantageous for businesses as it automates work by leveraging intelligent computers programmed with computer languages. If the goal is to achieve an intelligent link, the primary function of RPA is to implement AI (Connell and Brady, 1985).

RPA streamlines manual tasks, accelerating corporate growth by enabling real-time communication to identify appropriate candidates through job analysis and providing guidance on how to apply for the position. RPA will also assist recruiters in eliminating candidates who are not relevant to the posted vacancy (Aggarwal and Kaur, 2024). RPA bots inform candidates of their rejection, providing detailed explanations for the rejection and facilitating feedback on deficiencies in skills and qualifications. Additionally, RPA-enabled technology sends notifications and schedules interviews, while also communicating the specific skills that the organization is seeking for future recruitment. RPA technology empowers chatbot assistants to facilitate real-time communications, harness sentiments, and enhance interaction with candidates. Candidates engage with advanced RPA technology via many communication channels such as text messaging, email, social media, messaging apps like WhatsApp, Slack, Skype, and ATS (Applicant Tracking System). The RPA technology enhances the efficiency of the recruiting process by providing the recruiter with comprehensive information about the candidate, including their CV, cover letter, contact data, work experience, educational background, talents, reference checks, and driver's license. Additionally, it offers frequently asked questions (FAQs) related to job advertisements and interview scheduling. The advantages of RPA in enterprises are evident in its ability to automate routine tasks, cutting time to focus on important goals, accomplishments, and value-added initiatives for every stakeholder of business. Personnel would allocate their energies towards other activities that necessitate analytical reasoning and interpersonal abilities (Lacity and Willcocks, 2016).

According to Deloitte's human capital trends report, approximately 47% of HR leaders have accepted that most organizations are already engaged in technology-based HR practices. However, only 24% are currently utilizing AI and Robo-enabled technologies in their routine operations. Additionally, a LinkedIn survey stated that approximately 35% of hiring managers predict that by 2020, RPA will be one of the most important factors in the HRM industry, especially the ones engaged in recruitment processes (Lecaillon, Halpin, and Hennebert, 2018). RPA is shifting the facet of competitive digitalized economy and recruitment industry as it saves time, controls expenditure on talent acquisition, and elevates the gains and quick decision-making.

From the above, it can be drafted that AI and robotics are transforming HRM hiring practices and are creating an impact on the global economy.

1.3 METHODOLOGICAL RESEARCH

The article is purely a work of exploration and description. The study has been conducted on the basis of an extensive literature review and conceptual framework. The article is a work of secondary and qualitative data to analyze the importance and contribution of AI and Robotics in the hiring process and its benefits to the organizations. This chapter also tries to lay out the challenges that organizations may face while adopting and implementing AI and robotic-initiated hiring processes.

1.4 COMPANIES USING AI AND ROBOTIC PROCESS AUTOMATION (RPA) IN RECRUITMENT

Table 1.1 presents Indian companies and their green marketing initiatives.

TABLE 1.1 Indian Companies and their Green Marketing Initiatives

Sl. No	Company Name	AI or RPA Used in Hiring
1.	Electrolux (Home Appliances Brand)	• Uses AI in Hiring, with help of AI the recruiters are able to nurture leads by sending appropriate content to candidates based on their specified preferences, career goals, and hobbies. • Utilization of AI powered fit score of platforms, enhancement of candidate screening, matching and selection process to get more suitable candidates in a shorter time frame takes place. • When prospects apply and meet the requirements for an interview, an automatic AI scheduling system eliminates the need for time-consuming communication, therefore speeding up the process and improving the overall experience.
2.	IBM Watson Talent	• IBM Watson Tawlent's AI technology enhances efficiency and automates the recruitment process. This system identifies suitable candidates by analyzing data from social media, resume and other sources. • This system also provides chatbot that can respond to candidates' questions in tailored comments. The purpose of organization is to decrease the time-consuming process of recruitment into more cost and time effective process and also to amplify candidate involvement.
3.	Unilever	• The organization has created AI driven chatbot for responding to inquiries in a tailored manner. • It employs AI to scan the data and select the most suitable candidate for each position. Hence reducing time in recruitment process. In addition to this AI of the organization also points out discrepancies in employees and skill sets and provides them advise for future growth.
4.	Amazon	• Automated Review system created by company draws out comparable analysis of potential candidates with that of existing employees holding key positions, so as to identify such candidates who have more potential to achieve success. • Highly qualified applicants are selected for interviews, simplifying the recruitment procedure for professional office positions and distribution canters.

TABLE 1.1 *(Continued)*

Sl. No	Company Name	AI or RPA Used in Hiring
5.	Stanford Health Care	• AI chatbot used for recruitment by the company, simplifies the application process by allowing candidates to finish application by using cell phones if necessary. • Their chatbots helps in finding job matches, simplifies application process, communicates and answers FAQ's.
6.	Procter & Gamble (P&G)	• The companies AI driven-technologies helps in scanning dataset, resumes so forth, in order to find compatible matches for certain positions. Its AI algorithm assists in identifying the most suitable candidate as per the skill set, qualification and cultural compatibility required by organization. • AI-powered chatbots are used by company to enhance the interaction by successfully replying to queries, furnishing employee and organizational details and optimizing hiring process by enhancing employee experience.

Source: Author's Observation.

1.5 ROBOTIC PROCESS AUTOMATION TOOLS/COMPANIES

Table 1.2 lists companies developing and providing Robotic Process Automation (RPA) tools.

TABLE 1.2 Companies Developing and Providing RPA Tools

Sl. No	Company	RPA Tool
1.	UI Path	• UiPath offers a Software-as-a-Service (SaaS) solution using the Azure Cloud platform, coupled with built-in analytics and an electron application for its Orchestrator tool. – Notably, it includes an AI integration fabric that allows for the reliable implementation of AI features with robust capabilities. In order to expedite the process of implementation, the program incorporates a unified collection of pre-constructed automation elements to enhance customized automation. In 2019, UiPath purchased Process Gold and Step Shot, thereby enhancing its RPA functionality with advanced process mining capabilities.
2.	Automation Anywhere	• It is a cloud-based and SaaS-enabled platform that offers a cost-effective solution for small and midsize businesses (SMBs). It provides a wide range of features and a well-planned parameter that is appropriate for customers. • Automation Anywhere is renowned for its rapid implementation. The software offers developers the ability to integrate pluggable application programming interfaces (APIs). The company's electronic application enables clients to monitor and control bots while on the move. IQ Bot also offers handwritten text assistance now.

TABLE 1.2 *(Continued)*

Sl. No	Company	RPA Tool
3.	Edge Verve	• Edge Verve is a company that provides RPA for call centers. It caters to needs of large-scale organization dealing in customer-based services. It is a subsidiary of Infosys Ltd, an Indian origin software-based organization. • Apart from the above-mentioned RPA, company also provides ML and AI capable software's under the name Infosys Nia.
4.	Automation Edge	• Automation Edge RPA, with its AI capacity, may assist in the analysis of a company's data, providing the firm with enhanced insights into its systems. The utilization of Automation Edge RPA enables a company to achieve cost reduction and enhanced productivity by providing personalized services and self-service capabilities. Automation Edge RPA caters to the unique requirements of its customers and larger enterprises as they expand.
5.	Kofax	• Proficient in extracting data from documents. • It is possible to utilize APIs to extract and transfer data from several sources. • The ability to easily reuse and redeploy existing automations facilitates speedier deployments in new scenarios.

Source: Author's Observation.

1.6 USE OF AI AND ROBOTICS IN HIRING

Figure 1.3 illustrates three major steps in the AI- and robotics-enabled hiring process.

FIGURE 1.3 Showing three major steps followed by AI and robotic enabled hiring process.

Source: Author's observation.

1.6.1 CANDIDATE SOURCING

As per research from LinkedIn, 90% of professionals globally indicate a preference for recruiters to furnish details regarding potential positions within firms. A multitude of candidates aggressively pursue favorable employment opportunities, posing a difficulty for recruiters to effectively engage with exceptionally qualified individuals. RPA technology resolves this problem by automating the online sourcing process, systematically searching across many talent pools and databases to locate individuals of exceptional quality. By utilizing bots to actively search for people online, recruiters are able to free up more time to engage in consultations with clients regarding strategic talent initiatives, such as improving employment offers and bolstering marketing efforts.

1.6.2 ROBOTIC CHATBOTS

This application utilizes RPA technologies to construct chatbots that utilize sentiment analysis and computational linguistics to identify the candidate's speech and generate suitable responses. It facilitates a more collaborative, prompt, and engaging applicant experience by reducing the time between application submission and recruiter response. Chatbots carry out several functions, including requesting candidates to provide their resumes, cover letters, and contact information, inquiring about the candidates' education, experiences, and talents. They also address FAQs from candidates on job advertisements, the application process, and the hiring process. They also arrange interviews with hiring managers or recruiters.

1.6.3 HIRING CAMPAIGNS

RPA assists organizations by consistently engaging and connecting with applicants. The HR department's recruiters construct captivating email campaigns that can be strategically scheduled and automated to gradually deliver messages to both passive and active applicants, resulting in a significant level of engagement. The automated recruitment marketing campaigns consistently communicate and provide suitable job possibilities to different candidates, thereby enhancing the employer's brand and establishing a strong talent pool.

1.6.4 SCHEDULING INTERVIEWS

The integration of RPA technology into the interview scheduling process benefits both the HR team of enterprises and the candidates. This technology enables hiring managers or recruiters to offer candidates immediate access to the recruiting process and propose particular time slots for interviews based on availability. Once the candidates have chosen the interview time, an automated invitation is sent to both the interviewers and interviewees, blocking the scheduled time for additional candidates. When a business implements a video interviewing platform, automated bots send emails providing instructions on how to finish the video interview. After evaluating the interviews, the recruiters or hiring managers choose the most qualified individuals for the appropriate role.

1.6.5 ELEVATED CANDIDATE JOURNEY

The organization effectively manages the candidates' experiences using RPA solutions, which employ bots to gather real-time information on candidate experiences and assess their satisfaction levels throughout the recruitment process. Recruiters employ programmed bots to automatically produce email solicitations for feedback or brief surveys, which are used to track the candidates' progress during the hiring process. Automation enhances and oversees candidate experiences, while not every aspect of the process should be delegated to technology.

1.6.6 AFFINIX

Affinix should be employed in various roles associated with automation, such as RPA software, AI software, chatbots, and automated email drip campaigns.

1.7 ADVANTAGES OF COLLABORATING AI AND ROBOTICS IN HR

Use of AI and Robotics in HR practices has resulted in bringing laurels to organizations as:

1. Its cost-effective method of acquiring and generating pool of applicants.
2. It increased productivity.
3. It resulted in more efficient talent acquisition.

4. It reduced employee churn, making sure employees remain in organization for longer time period, by analyzing their behavior patterns and helping in quick decision making.
5. It increased accuracy in HRM Practices.
6. It is time saving, as automation fastens the process.
7. It is low risked and is easy to incorporate.
8. It creates new opportunities for employees, by preparing HR for new trends and innovations, leading demand for professionals in field of AI and Robotics.

1.8 CHALLENGES IN USING AI AND ROBOTICS IN HR

Use of AI and Robotics in HR practices is not that easy as it seems, it comes with following limitations/challenges faced by organizations:

1. Occasionally, the data obtained using AI may contain irrelevant or useless information, which further complicates the process for HR when they are unable to find an exact match.
2. AI is unable to effectively categorize the profiles for high-level positions.
3. AI is based on binary logic, which means it relies on a system of ones and zeros. However, HR departments do not consist solely of individuals who are completely good or completely bad. As a result, it can often be challenging to create a work environment that is conducive to employee comfort.
4. Another challenge of using AI in hiring process is that of personal privacy and the handling and analysis of data. Both HR experts and online HRM users are concerned with analyzing data and sharing their own information.
5. Resentment from employees as it will lead to loss of jobs by replacing HR.

Hence above are the challenges or issues faced by organizations, while adopting new technologies replacing HR.

1.9 CONCLUSION AND FUTURISTIC APPROACH

From the above extensive explanation, it can be safely concluded that the inculcation of AI and robotics has transformed HRM practices completely.

The advancement of technology has led to better, more accurate, and quicker decision-making and talent acquisition for organizations. It has further advanced the way of doing business, making it cost- and time-effective. It has eased the process of hiring by effective CV screening, interview scheduling, so forth, hence accelerating the overall hiring process. AI has expanded its scope beyond recruitment and selection, currently aiding HR teams in performance management and payment processes.

Moreover, the evolution of RPA technology has further provided organizations with the opportunity for elevated productivity by reducing time, cost, and unnecessary HR. RPA technology is a crucial tool for optimizing the recruitment and hiring process. The implementation of technology in the recruiting process has facilitated talent acquisition and hiring professionals in making faster and more accurate decisions by enhancing candidate efficiency throughout the hiring process. Hence, the implementation of RPA technology has a substantial influence on an organization's ability to achieve optimal equilibrium between skilled personnel and human engagement.

Both Robotics and AI have held hands in shifting the paradigms and processes of talent acquisition. Although these technologies are not free from limitations or challenges, one of the most obvious reasons being resentment among employees due to job loss at the hands of technological advancements, adding to the pressures of unemployment, the positive impacts of these technologies still outrun the negative impacts, as they are also providing job opportunities for AI and Robotic professionals due to increased demand.

KEYWORDS

- **artificial intelligence**
- **automation**
- **human resource management**
- **recruitment technology**
- **robo-advisors**
- **robotic process automation**
- **talent acquisition**

REFERENCES

Abdullah, H. (2009). Evolving terms of human resource management and development. *Journal of International Social Research, 2*(9), 180–186.

Acikgoz, Y. (2019). Employee recruitment and job search: Towards a multi-level integration. *Human Resource Management Review, 29*, 1–13.

Aggarwal, K. K., & Agrawal, S. (2024). Artificial intelligence and its role in financial market. In *Global Financial Analytics and Business Forecasting.*

Aggarwal, K. K., & Kaur, R. (2024). Transforming people management: Harnessing HR analytics in the transhuman era. In *Embracing Transhumanism and Genomics in Human Resources Management* (pp. 117–134). IGI Global.

Aggarwal, K. K., & Singhal, S. (2024). The effect of perceived philanthropic activities of corporate organizations on consumer's purchase intention. *International Journal of Business and Globalization, 36*(2, 3), 212–224. https://doi.org/10.1504/IJBG.2024.137401 (accessed on 11th August 2025).

Aggarwal, K. K., & Taneja, D. (2024). Impact of organizational culture on employee satisfaction, motivation, performance, commitment, and engagement in educational institutions. In *Building Sustainable Human Resources Management Practices for Businesses*. IGI Global. https://doi.org/10.4018/979-8-3693-1994-9.ch012 (accessed on 11th August 2025).

Andersson, G., Hallén, N., & Smith, P. (2016). *Recruitment and Selection—Theory and Practice* (1st ed., pp. 1–83). Lund: Studentlitteratur AB.

Bas, A. (2012). Strategic HR management: Strategy facilitation process by HR. *8th International Strategic Management Conference*, 313–321.

Biswas, N. (2012). Recruitment and selection process at Viyellatex group.

Chapman, D. S., & Gødøllei, A. F. (2017). *The Wiley Blackwell Handbook of the Psychology of the Internet at Work.* Newark, United Kingdom: John Wiley & Sons, Incorporated. http://ebookcentral.proquest.com/lib/uisbib/detail.action?docID=4987368 (accessed on 11th August 2025).

Chenik, M. (2018). Artificial intelligent. AI. Stockholm: Karolinska Institutet. https://ki.se/sites/default/files/2018/04/27/artificiell_intelligens.pdf (accessed on 11th August 2025).

Connell, J. H., & Brady, M. (1985). Learning shape descriptions. In *Proceedings of the 9th International Joint Conference on Artificial Intelligence* (Vol. 2, pp. 922–925).

Deloitte. (2019). Leading the social enterprise: Reinvent with a human focus. Deloitte. https://www2.deloitte.com/content/dam/Deloitte/cz/Documents/humancapital/cz-hc-trends-reinvent-with-human-focus.pdf (accessed on 11th August 2025).

Dhamija, P. (2012). E-recruitment: A roadmap towards e-human resource management. *Researchers World, 3*(3), 33.

Faliagka, E., Ramantas, K., Tsakalidis, A., & Tzimas, G. (2012). Application of machine learning algorithms to an online recruitment system. In *Proceedings of the International Conference on Internet and Web Applications and Services.*

Garg, A., Gaur, S., & Sharma, P. (2023). A review paper: Role of artificial intelligence in recruitment process. https://www.conclusion.nl/nieuws/partnership-another-monday (accessed on 11th August 2025).

Griepentrog, B. K., Harold, C. M., Holtz, B. C., Klimoski, R. J., & Marsh, S. M. (2012). Integral social identity and the theory of planned behavior: Predicting withdrawal from an organization recruitment process. *Personnel Psychology, 65*, 723–753.

HireVue. (2017). HireVue: Case study. https://cdn2.hubspot.net/hubfs/464889/Hilton%20Aug%202017/2017_12_SuccessStory_Hilton_CustomerMarketing3.pdf?__hstc=&__hssc=&hsCtaTracking=b76cefe9-cece-4631-bef5-53084aa900e5%7Ca26c5bca-5fbe-46e3-b0e4c1f79894f72c (accessed on 11th August 2025).

HireVue. (2018). HireVue. https://www.hirevue.com/products/video-interviewing (accessed on 11th August 2025).

Holm, A. B. (2012). E-recruitment: Towards a ubiquitous recruitment process and candidate relationship management. Aarhus V: Rainer Hampp Verlag.

Jatoba, M., Gutierriz, I., Fernandes, P. O., Teixeira, J. P., & Moscon, D. (2019). Artificial intelligence in the recruitment & selection: Innovation and impacts for the human resources management. In *Economic and Social Development: Book of Proceedings* (pp. 96–104).

Johnson, R. D., Stone, D. L., & Lukaszewski, K. M. (2020). The benefits of eHRM and AI for talent acquisition. *Journal of Tourism Futures*. Advance online publication.

Kaczmarek, T., Kowalkiewicz, M., & Piskorski, J. (2005). Information extraction from CV. In *Proceedings of the 8th International Conference on Business Information Systems* (pp. 3–7).

Kaplan, A., & Haenlein, M. (2019). Siri, Siri, in my hand: Who's the fairest in the land? On the interpretations, illustrations, and implications of artificial intelligence. *Business Horizons, 62*(1), 15–25.

Kaur, R., & Aggarwal, K. K. (2024). Navigating the future: A comprehensive study on robo-advisors in wealth management. In *Embracing Transhumanism and Genomics in Human Resources Management* (pp. 247–260). IGI Global.

Kok, J. D., & Uhlaner, L. M. (2011). Organization context and human resource management in the small firm. *Small Business Economics, 17*(4), 273–291.

Komakech, E., Obici, G., & Mwesigwa, D. (2021). Efficacy of inspirational motivation on the performance of public health workers in mid-north of Uganda. *Annals of Human Resource Management Research (AHRMR), 1*(2), 85–97.

Lacity, M. C., & Willcocks, L. P. (2016). A new approach to automating services. *MIT Sloan Management Review, 58*(1), 41–49.

Lecaillon, B., Halpin, L., & Hennebert, N. (2018). Getting started agenda here with you today RPA today—RPA tomorrow: What are the challenges and insights from the expert.

Lee, I. (2005). The evolution of e-recruiting: A content analysis of Fortune 100 career web sites. *Journal of Electronic Commerce in Organizations, 3*(3), 57–68. https://doi.org/10.4018/jeco.2005070104 (accessed on 11th August 2025).

Lisa, A. K., & Talla Simo, V. R. (2021). An in-depth study on the stages of AI in recruitment process of HRM and attitudes of recruiters and recruitees towards AI in Sweden (Dissertation). https://urn.kb.se/resolve?urn=urn:nbn:se:umu:diva-184521 (accessed on 11th August 2025).

Loucks, J., Hupfer, S., Jarvis, D., & Murphy, T. (2019). Future in the balance? How countries are pursuing an AI advantage. *Deloitte Insights*, 1–24.

Madakam, S., Holmukhe, R. M., & Jaiswal, D. K. (2019). The future digital workforce: Robotic process automation (RPA). *Journal of Information Systems and Technology*

Management, 16, 1–16. https://doi.org/10.4301/s1807-1775201916001 (accessed on 11th August 2025).
Maini, V., & Sabri, S. (2017). Machine learning for humans. *Medium*. https://medium.com/machine-learning-for-humans/why-machinelearning-matters6164faf1df12 (accessed on 11th August 2025).
McCarthy, J. (1998). What is artificial intelligence?
Microsoft. (2018). Artificial intelligence in Europe: How 277 major companies benefit from AI—Outlook for 2019 and beyond. Microsoft. https://news.microsoft.com/uploads/prod/sites/153/2018/09/AI-report-Sweden.pdf (accessed on 11th August 2025).
O'Donovan, D. (2019). HRM in the organization: An overview. In *Management Science: Management and Industrial Engineering* (pp. 75–110).
Rąb-Kettler, K., & Lehnervp, B. (2019). Recruitment in the times of machine learning. *Management Systems in Production Engineering, 27*(2), 105–109.
Raub, M. (2018). Bots, bias and big data: Artificial intelligence, algorithmic bias and disparate impact liability in hiring practices. *Arkansas Law Review, 71*(2), 529–570.
Russell, S., & Norvig, P. (2009). *Artificial Intelligence: A Modern Approach* (3rd ed.). Saddle River, NJ: Prentice Hall.
Salomon, G., Perkins, D. N., & Globerson, T. (1991). Partners in cognition: Extending human intelligence with intelligent technologies. *Educational Researcher, 20*(3), 2–9.
Sekhri, A., & Cheema, J. (2019). The new era of HRM: AI reinventing HRM functions. *International Journal of Scientific Research and Review, 7*(3), 3073–3076.
Sharma, A. (2018). How AI reinvented hiring practice at L'Oreal. *People Matters*. https://www.peoplematters.in/article/technology/how-the-worldslargest-cosmetic-company-transformed-its-hiring-practice-with-ai-19006 (accessed on 11th August 2025).
Strohmeier, S., & Parry, E. (2014). HRM in the digital age: Digital changes and challenges of the HR profession. *Employee Relations, 36*. https://doi.org/10.1108/ER-03-2014-0032 (accessed on 11th August 2025).
Tecuci, G. (2012). Artificial intelligence. *Wires Computational Statistics, 4*(2), 168–180.
Upadhyay, A. K., & Khandelwal, K. (2018). Applying artificial intelligence: Implications for recruitment. *Strategic HR Review, 17*(5), 255–258.
Ved, S., Kaundanya, N. S., & Panda, O. P. (2016). Applications and current achievements in the field of artificial intelligence. *Imperial Journal of Interdisciplinary Research, 2*(11), 932–936.
Wall, T. D., & Wood, S. J. (2005). The romance of human resource management and business performance, and the case for big science. *Human Relations, 58*(4), 429–462.

CHAPTER 2

Robo-Advisor: Augmenting Quality of Work-Life and Employee Well-Being in Industry 5.0

AMRISHA MINOCHA,[1,*] GIRISH LAKHERA,[2] SHAVETA BHATINA,[3] and BHUPINDER PAL SINGH CHAHAL[4]

[1]*Amity University, Noida, Uttar Pradesh, India*

[2]*Department of Management Studies, Graphic Era (Deemed to be University), Dehradun, Uttarakhand, India*

[3]*Center for Distance and Online Education, Manav Rachna International Institute of Research and Studies, Faridabad, Haryana, India*

[4]*Yorkville University, British Columbia, Canada*

*Corresponding author

ABSTRACT

The ascendancy of Industry 5.0 epitomizes a paradigmatic shift wherein intelligent machines evolve from being mere tools of automation to collaborative partners in augmenting human capabilities. Within this context, robo-advisors have emerged as pivotal agents in transforming organizational practices, particularly by enhancing quality of work life (QWL) and fostering employee well-being. Unlike their conventional application in financial services, robo-advisors are increasingly integrated into human resource ecosystems, where they automate repetitive tasks, streamline decision-making, and offer personalized guidance to employees. This chapter interrogates the intersection of robo-advisory technologies, human–robot interaction, and psychosocial dynamics of the workplace,

Robo-Advisors and Artificial Intelligence in Human Resources Management: Revolutionizing HR.
Sanjay Taneja, Swati Gupta, Girish Lakhera, Mohit Kukreti, and Ercan Özen (Eds.)

DOI: 10.1201/9781779641410-2

emphasizing their capacity to mitigate occupational stress, reduce attrition, and facilitate a more sustainable work–life balance. Anchored in an extensive review of extant literature, the study reveals that robo-advisors not only accelerate efficiency but also catalyze the evolution of "emotional AI," which aspires to discern affective states and pre-empt burnout. However, the inquiry simultaneously illuminates the dialectics of opportunity and peril: while these technologies engender empowerment and inclusivity, they equally provoke ethical dilemmas, data privacy concerns, and anxieties surrounding human obsolescence. Ultimately, the chapter posits robo-advisors as both disruptors and enablers in the architecture of Industry 5.0, redefining the epistemology of work and well-being.

2.1 INTRODUCTION

The primary focus of industrial revolutions is often centered around the distinction between human labor and machine operations. Whether referred to as a "machine" or a "robot," these automated systems are expected to gradually assume tasks that are challenging, monotonous, or pose potential risks, relieving humans of such responsibilities. The concept has evolved over time, starting from the first industrial revolution, which initially focused on harnessing power from water and steam. It has since progressed to include advanced technologies such as Artificial Intelligence (AI) and the Internet of Things (IoT), constituting what is now known as Industry 4.0 (Demir, Döven, and Sezen, 2019). But now researchers are looking forward to the fifth industrial revolution (also called robot-human co-working) where the basic idea is that machines will not work for us but with us (Ganer et al., 2022). Therefore, currently, efforts are being made to develop Industry 5.0 through a more consistent management of human resources (HR) (Santana and Cobo, 2020).

AI bots and humanoid service robots are garnering more and more interest from the industry (Larivière et al., 2017; Vrontis et al., 2022; Zhang, Pentina, and Fan, 2021; Chi et al., 2023. Nowadays, a wide range of industries use robots for repetitive tasks like medical (Srivastava et al., 2016), education (Mubin et al., 2013), social robots (Lopes, Ferreira, and Prada, 2023), and even the banking and investment industry (Alsabah et al., 2021). Robo-advisors, also spelled Robo-adviser or Robo advisor, are digital platforms offering automated financial planning services driven by algorithms, often requiring minimal human supervision. Robo-advisors are online platforms that automate and improve investing procedures by using machine learning (ML) and AI algorithms. These platforms collect critical data from investors, including time horizon, financial goals, and risk tolerance, and use that information to build diversified portfolios that are customized to meet each investor's needs. Therefore, the banking industry has been the pioneer in adopting the concept of Robo-advising to make important decisions regarding investments and portfolio management (Shanmuganathan, 2020). The Robo advisory market has experienced exceptional growth over the last several years. The market's estimated size as of 2023 was USD 6.61 billion. From 2024 to 2030, it is anticipated to grow at a compound annual growth rate (CAGR) of 30.5%.

The HR department has now also tried to embed the use of Robo advisors at various HR levels, for example, through the introduction of E-HRM, which offers numerous opportunities to reduce the costs associated with HRM operations, for example, in the assessment of job candidates and employee performance (Cooke et al., 2019). Likewise, human-robot interaction (HRI) in robotics is becoming increasingly important, has attracted the interest of several disciplines, and holds significant potential for personnel departments too. Traditional human-operated manual systems and fully automated robotic systems are where HRI occurs (Bruno and Antonelli, 2018). However, the study of employee well-being and quality of working life is still in its infancy.

Multiple studies have been conducted on employee well-being and quality of work-life in silos, but not in context to each other through the lens of Robo-advising in the academic world. Therefore, the aim of this chapter is to deep dive into the depths of understanding the evolution of Robo-advisors and elaborate on the modern definition of quality of work-life and employee well-being in Industry 5.0. Also, how augmenting/introducing Robo-advisors has already improved the above two, and how they intend to make it even better in the upcoming future. The chapter will also shed light on the organizations using Robo-advisors in their HR administration.

2.2 LITERATURE REVIEW

2.2.1 ROBO ADVISORS IN INDUSTRY 5.0

Industrial upcycling was the main concept of a LinkedIn article titled "Industry 5.0 – from virtual to physical," which introduced the term Industry 5.0 in 2015. The aim of the article was to convey that the world should move from Industry 4.0 to Industry 5.0, where machines are not our servants, but our partners. In the research paper entitled "Industry 4.0 and Industry 5.0 – Emergence, Conception and Perception" (Xu et al., 2021), the author tries to explain the connection between Industry 4.0 and 5.0, since Industry 5.0 is dependent on Industry 4.0 in terms of its technological innovations and the social aspects of production, social responsibility, and value systems, so forth, since Industry 5.0 is purely value-oriented. Therefore, it can be said that Industry 5.0 is a "human-centric solution," while Industry 4.0 is still focused on automation. Demir (2019) highlighted the human-robot coworking vision in Industry 5.0 in his research. A recent LinkedIn article (Christian, 2022) stated that Industry 5.0 is all about how people can collaborate with robots and IoT devices in the automated industrial environments of the future. Industry 5.0 emphasizes the human aspect and how new technologies such as IoT and big data can improve human skills and abilities, whereas Industry 4.0 focuses on the usage of robotic systems and other intelligent machines to enhance both quality and optimize productivity in the manufacturing domain.

Since Industry 5.0 is being referred to as "human-robotic interaction," the importance of Robo-advisors is also being recognized. Although scholars disagree on the degree of rigor in the three components, IEEE (Guizzo, 2018) defines a robot as an autonomous device that perceives, processes, and acts in reality. Another study by Gan, Khan, and Liew (2021) stated that Robo-advising has three variables, namely trust, financial knowledge, and tendency to rely on a Robo-advisor. It was discovered that only performance expectancy and social influence are the primary construct variables that are significant. Furthermore, other factors that affect the intention to use Robo-advisors include perceived financial knowledge, trust factors, and a propensity to rely on them.

However, there have been a couple of studies in the past that have talked about the bane of having a robot advisor in the organization. These researches highlighted the issues pertaining to the evolution of organizational behavior, work ethics, trust/authenticity, and privacy linked to the Robo advisors and acceptance of Robo-advisors (Demir, Caymaz, and Elçi, 2017).

2.2.2 QUALITY OF WORK LIFE

Employee satisfaction in terms of their personal requirements and work experience that they derive from the organization that they work with is referred to as quality of work life, or QWL (Impas, 2018).

Employee well-being and medical care, various employee incentive programs, work security, workplace sufficiency, workplace design, minimizing conflict, participation in management decisions, training and reward systems, and more all have an impact on how happy our employees are in the workplace (Saleem and Malik, 2023; Sahni, 2019).

Quality of work life is the most anticipated and addressed issue, either directly or indirectly, to assess and determine its explicit or implicit impact on employees in the organization (Leitão, Pereira, and Gonçalves, 2019). It was only when employers realized its importance that they began to appreciate the role of AI in improving the quality of work life of their employees. They introduced AI techniques and tools such as Robo Advisor and other robotic technologies in various HR functions. These consultants were tasked with performing routine and repetitive tasks so that employees could learn more skills relevant to their jobs.

Meanwhile, the contemporary concept of quality of working life is evolving to integrate elements of AI and robotics (Majumder and Biswas, 2021). Robots are changing the nature of work in the future and are already having a major impact on workplace dynamics. The level of integration of AI into the workplace for both employers and employees is examined in a recent study, namely UKG, an American multinational technology company. Surveys were conducted among over 4000 workers in nine countries, including India, to understand their thoughts on the changing role and adoption of robots (AIT News Desk, 2023).

Studies show that 72% of Indian workers directly interact with AI-powered workplace tools, such as Robo-advisors, systems, and procedures, in their companies, and 75% of them use generative AI tools for both professional and personal use, such as ChatGPT (75%), OpenAI (47%), and Google Bard (22%) (AIT News Desk, 2023). All of these tools, as a whole, will only have a positive impact on improving employees' working lives if they are used together with the individual and not by replacing him. Therefore, the quality of working life can now be described as a "movement" with both international and technical influence (Warhurst, Mathieu, and Dwyer, 2022).

2.2.3 EMPLOYEE WELLBEING

The global confront of 2020 is starting to embrace employee well-being as a tenet. In every organization, the idea of fostering organizational growth is becoming more and more respected. In 2010, the British Psychological

Society acknowledged the significance of employee psychological well-being and its influence on organizational success.

Researchers have echoed a lot of definitions about well-being in their studies, covering various aspects and dimensions that characterize it. As per Russell and Daniels (2000), some of the affective, behavioral, and cognitive components that make up psychological well-being at work include competence, autonomy, integration of information, and the capacity to feel both positive and negative emotions.

The concept of well-being is multifaceted and intricate, and it has been defined and operationalized in several inconsistent ways. Fisher (2014) viewed well-being as both hedonic and eudaimonic, where hedonic meant temporary pleasant feeling and eudaimonic meant that autonomous development and inner growth are components of self-actualization. Some studies have echoed the concept of employee well-being by broadly defining it into psychological well-being and physical well-being, where the former is "subjective experience of an individual" and the latter is "bodily health and functioning" (Grant, Christianson, and Price, 2007). Warr, Sinclair, and Tetrick (2012) stated that psychological well-being can be classified into affective and cognitive processes. Transition in the working processes, tasks, structures, and hierarchies at the workplace has occurred due to the introduction of new technologies and AI techniques.

The notion of worker well-being is multifaceted, encompassing both objective factors like pay and benefits and subjective ones like motivation, contentment, social connection, and significance. Businesses and their HR departments long ago realized that these aspects of working life had an impact on employees' productivity and turnover (Warr, Sinclair, and Tetrick, 2012). On the contrary, groundbreaking research by Giuntella and Stella (2023) also corroborated a divergence in life satisfaction levels among exposed workers since the notable increase in AI adoption in 2015. Among workers exposed to AI, job satisfaction has significantly decreased, and concerns regarding job security and personal financial circumstances have increased.

2.2.4 ROBO-ADVISING IN IMPROVING QWL AND EWB

AI and robotics are key to overcoming modern workforce management challenges and creating a more fulfilling work experience, thus balancing work-life. There are a couple of ways through which Robo-advisors aim to overcome modern workplace challenges, like enhancement of work productivity by preventing schedule changes, minimizing turnover by giving employees flexibility to work through both hybrid and remote modes, automation of repetitive tasks, and utilization of skills.

Robots and Robo-advisors are positively influencing workers by creating new work opportunities (Zhou, 2017) and contributing by inflating the income of managerial positions, having positive impacts on workers, like producing new employment opportunities and increasing the income of some managerial and specialist positions (Xu, Xue, and Zhao, 2023). The transactional stress theory suggests that people use problem-focused coping mechanisms to reduce stress and increase long-term well-being when they believe that AI offers opportunities for them (Folkman et al., 1986). According to Penley, Tomaka, and Wiebe, people's physical and mental well-being also improves when they use problem-focused problem-solving techniques to cope with workplace stressors (Penley, Tomaka, and Wiebe, 2002).

Companies that integrated robot advising found that their workweeks were shortened, allowing their staff to take longer vacations. A lot of pioneering research is gaining attention on "Emotional AI" as well, which comprises AI tools and robots that are designed to sense the affective behavior of the employees, like happiness, anger, alertness, agitation, nervousness, surprise, dissatisfaction, so forth, (Ho, Mantello, and Ho, 2023). Thus, Robo advisors, also in the form of IoT or wearables, are opening avenues for both quality of work life and employee wellbeing. It advises them to manage their stress. A lot of this suggests that integrating robots into the workplace improves work-life balance and gives people the chance to improve their personal lives as well.

Some newly designed robots are also advising organizations on the burnout of employees and suggesting measures to prevent it. Some highly reputed journals published research in the *Journal of Occupational Health Psychology* that revealed that robots were highly accurate at predicting burnout. AI tools can assist managers and HR specialists in identifying staff members who may be experiencing extreme stress or exhaustion by analyzing a variety of data points, including work hours, workload, and even communication patterns. Organizations can take proactive steps to prevent burnout, like adjusting the workload, offering flexible work schedules, or providing more support, thanks to this early detection (Proactive. IT Appointments Ltd., 2023).

2.3 LATEST ROBO-ADVISOR TECHNOLOGIES USED BY ORGANIZATIONS

Using a predetermined set of skills and a comparison of employee skill sets with those of external workers performing similar roles, Johnson and Johnson uses robots to assess employee skill levels. The organization can detect skills gaps and take appropriate action to close them thanks to this methodology. Likewise, AI-powered chatbots can conduct initial interviews and respond to FAQs, making the onboarding process for new employees seamless. This quickens the hiring process and ensures that qualified applicants are matched with suitable positions, which eventually raises employee satisfaction and retention rates.

Betterment met with Charles Schwab, a giant in the financial services business, for their employees with the help of this Robo-advisory collaboration. Betterment also offers some educational materials so that employees can learn about the various investment options and make more informed decisions. It gathers data from employees through questionnaires to determine their investment risk profile and help them find the best strategy based on individual objectives and appetite for risk.

Blooom partners with Willis Towers Watson-backed retirement plans. Working employees can then select retirement goals in the platform, and Blooom will recommend specific investments designed to get them there by their planned age of retirement. The automation behind Blooom liberates Willis Towers Watson's HR team from having to manually run employee investment accounts, so HR can deploy their resources on other employee well-being initiatives or communications driving higher workforce participation. Acorns, a micro-investing app, similarly leans on Finhabits, also led by Laibow, and the platform behind Acorn's bite-sized financial literacy modules used with employees. These modules include topics on budgeting, savings, and credit cards.

Starbucks is improving employee wellbeing by utilizing advisory and sophisticated automation technologies. These systems are well-equipped to handle inventory control, staffing requirements forecasting, scheduling, and other related tasks. Furthermore, this kind of technology, precisely Robo-advisors, are capable of forecasting and maintaining the requirements for store equipment, thus making day-to-day working easy and smooth for its employees.

Additionally, AI-powered solutions like robo-advising can evaluate engagement metrics and employee sentiment, giving employers insightful information. This data can be used by organizations to identify and address issues related to work-life balance, employee morale, and overall job satisfaction. Businesses can create a win-win situation for their workers.

The company's Pepper robot has been used by many companies to perform customer service and greeting tasks associated with human resources (HR). In a similar vein, Hanson Robotics' humanoid robots, Sophia, are employed by many companies to perform HR functions like recruiting and orienting new employees, improving their bottom line by utilizing AI to support employee well-being and satisfaction.

Mahindra Group designed and developed Mahindra Robot, which is now efficient in resumé screening and candidate interviews, and is worth USD 19.4 billion. Romanian software UiPath is efficiently taking care of the RPA in the organization. Such advanced software cum Robo-advising software will now be able to help automate a variety of HR tasks, including the rudimentary processing of benefits and payroll.

Programs and robots driven by AI can track an individual's stress levels by analyzing data from wearables, surveys, or workplace communication platforms. With this information, AI can offer personalized recommendations for stress-reduction techniques, time management strategies, and even mental health services. With this customized approach, employees may be able to manage their own health. Furthermore, AI can enhance employee benefits programs and well-being. AI can provide insightful information about the best programs or resources by analyzing employee feedback and usage patterns. Employers can use this data-driven approach to customize their benefits and well-being programs to the specific requirements of their workforce.

On the contrary, no case study was published by Einola, Khoreva, and Tienari (2023) about an anthropomorphized robot named Max in a media consulting company. The research suggests that collective affect plays a crucial role in today's technology-driven organizations, where the role of people is often overlooked in the context of the current wave of AI technologies. Overzealous managerial enthusiasm for technology, coupled with a poor understanding of organizational life and a devaluation of the skills and knowledge that only humans can possess to complete tasks, are some of the reasons for unreasonable expectations of the performance of a particular AI solution.

Robots can significantly improve a healthier and more supportive work environment by providing personalized support, 24/7 support via chatbots, predictive analytics to prevent burnout, and improved employee benefits. AI can help workers deal with stress, anxiety, and other mental health issues with the right approaches and ethical considerations, ultimately leading to happier and more productive workplaces.

2.4 CONCLUSION

In 2016, research on Robo advisors started evolving; likewise, human-robot collaboration is the new theme in today's time since it is being studied in various other industries like the hotel and hospitality industry as well (Choi et al., 2020). To date, numerous researches have also been conducted in the field of finance and FinTech (Financial Technology) industry, where new themes like digital finance and innovation in digital finance, so forth, (Zavolokina, Dolata, and Schwabe, 2016) were studied. But in social science research, robot advisors remain a new concept, yielding a lot of opportunities. Therefore, the presented topic is relatively new and has great potential for growth in the upcoming future.

Some research in the past has conducted studies showing that HR departments do not motivate their employees well in using robots due to a lot of apprehensions. Due to this, robot collaboration and employees' behavioral intention regarding robot use are related, but HRM involvement was unable to moderate this relationship, and thus the HR department needs to change its strategies and policies regarding the same (Vrontis et al., 2022; Xu, Stienmetz, and Ashton, 2020). This can only happen when the department plans for the change strategically and prepares for such dynamic technological change.

The direct relationship of the impact of Robo advisors and robotic interventions on quality of work life and employee well-being has not been addressed to a larger extent to date in the research conducted thus far. Therefore, we are unsure of the extent to which automation will affect the quality of work-life. However, we argue that the qualitative changes that automation will bring to work and workspaces should be included in the discussion of automation and the future of work.

However, Robo advisors remain at an infancy stage in addressing affective AI, or Emotional AI. Enabling artificial systems to detect, comprehend, and react to human emotions is the main goal of the quickly evolving field of affective AI, or emotional AI. To fully achieve the promise of affective AI, several obstacles must be overcome. Being able to recognize and interpret human emotions with accuracy and consistency is one of the biggest challenges. As mentioned earlier, quality of work life is a sum of a person's experience about his/her job, where he goes through hundreds of emotions every day in daily work. Therefore, it cannot be accurately stated that a Robo advisor will not be able to figure out affective behavior of humans accurately.

2.5 MANAGERIAL IMPLICATIONS

We recommend through this study that new automation initiatives must incorporate elements of job design and employee experience. This way, automation will not always have unintended consequences for work-life quality, as these projects tend to focus only on the tasks that need to be automated.

KEYWORDS

- **automation**
- **HRM**
- **industry 5.0**
- **robo-advisors**
- **well-being**

REFERENCES

AIT News Desk. (2023). UKG survey: 95 percent of Indian employees believe AI can improve their quality of life. https://aithority.com/machine-learning/ukg-survey-95-percent-of-indian-employees-believe-ai-can-improve-their-quality-of-life/ (accessed on 11th August 2025).

Alsabah, H., Capponi, A., Ruiz Lacedelli, O., & Stern, M. (2021). Robo-advising: Learning investors' risk preferences via portfolio choices. *Journal of Financial Econometrics, 19*(2), 369–392. https://doi.org/10.1093/jjfinec/nbz040 (accessed on 11th August 2025).

Bruno, G., & Antonelli, D. (2018). Dynamic task classification and assignment for the management of human-robot collaborative teams in work cells. *The International Journal of Advanced Manufacturing Technology, 98*(9–12), 2415–2427. https://doi.org/10.1007/s00170-018-2400-4 (accessed on 11th August 2025).

Chi, O. H., Chi, C. G., Gursoy, D., & Nunkoo, R. (2023). Customers' acceptance of artificially intelligent service robots: The influence of trust and culture. *International Journal of Information Management, 70*, 102623. https://doi.org/10.1016/j.ijinfomgt.2023.102623 (accessed on 11th August 2025).

Choi, Y., Choi, M., Oh, M. (Moon), & Kim, S. (Sam). (2020). Service robots in hotels: Understanding the service quality perceptions of human-robot interaction. *Journal of Hospitality Marketing & Management, 29*(6), 613–635. https://doi.org/10.1080/19368623.2020.1703871 (accessed on 11th August 2025).

Christian, B., (2022). *What is Industry 5.0?* https://www.clarify.io/learn/industry-5-0#:~:text=As%20we%20said%2C%20Industry%205.0 (accessed on 11th August 2025).

Cooke, F. L., Liu, M., Liu, L. A., & Chen, C. C. (2019). Human resource management and industrial relations in multinational corporations in and from China: Challenges and new insights. *Human Resource Management, 58*(5), 455–471. https://doi.org/10.1002/hrm.21986 (accessed on 11th August 2025).

Cutrona, C. E., Russell, D. W., Hessling, R. M., Brown, P. A., & Murry, V. (2000). Direct and moderating effects of community context on the psychological well-being of African American women. *Journal of Personality and Social Psychology, 79*(6), 1088–1101. https://doi.org/10.1037/0022-3514.79.6.1088 (accessed on 11th August 2025).

Demir, K. A., Caymaz, E., & Elçi, M. (2017). Issues in integrating robots into organizations. *The 12th International Scientific Conference "Defense Resources Management in the 21st Century."*

Demir, K. A., Döven, G., & Sezen, B. (2019). Industry 5.0 and human-robot co-working. *Procedia Computer Science, 158*, 688–695. https://doi.org/10.1016/j.procs.2019.09.104 (accessed on 11th August 2025).

Einola, K., Khoreva, V., & Tienari, J. (2023). A colleague named Max: A critical inquiry into affects when an anthropomorphized AI (ro)bot enters the workplace. *Human Relations.* https://doi.org/10.1177/00187267231206328 (accessed on 11th August 2025).

Fisher, C. D. (2014). Conceptualizing and measuring wellbeing at work. In C. L. Cooper (Ed.), *Wiley*. https://doi.org/10.1002/9781118539415 (accessed on 11th August 2025).

Folkman, S., Lazarus, R. S., Gruen, R. J., & DeLongis, A. (1986). Appraisal, coping, health status, and psychological symptoms. *Journal of Personality and Social Psychology, 50*(3), 571–579. https://doi.org/10.1037/0022-3514.50.3.571 (accessed on 11th August 2025).

Gan, L. Y., Khan, M. T. I., & Liew, T. W. (2021). Understanding consumer's adoption of financial Robo-advisors at the outbreak of the COVID-19 crisis in Malaysia. *Financial Planning Review, 4*(3). https://doi.org/10.1002/cfp2.1127 (accessed on 11th August 2025).

Ganer, S. D., Kediya, S. O., Suchak, A. K., Dey, S. K., & Band, G. (2022). Analytical study of HRM practices in industry 5.0. *IOP Conference Series: Materials Science and Engineering, 1259*(1), 012041. https://doi.org/10.1088/1757-899X/1259/1/012041 (accessed on 11th August 2025).

Giuntella, O., Koenig, J., & Stella, L. (2023). Artificial intelligence and workers' well-being. *SSRN Electronic Journal.* https://doi.org/10.2139/ssrn.4589223 (accessed on 11th August 2025).

Grant, A. M., Christianson, M. K., & Price, R. H. (2007). Happiness, health, or relationships? Managerial practices and employee well-being tradeoffs. *Academy of Management Perspectives, 21*(3), 51–63. https://doi.org/10.5465/amp.2007.26421238 (accessed on 11th August 2025).

Guizzo, E. (2018). Your next t-shirt will be made by a robot. *IEEE Spectrum, 55*(1), 50–57. https://doi.org/10.1109/MSPEC.2018.8241738 (accessed on 11th August 2025).

Ho, M. T., Mantello, P., & Ho, M. T. (2023). An analytical framework for studying attitude towards emotional AI: The three-pronged approach. *MethodsX, 10*, 102149. https://doi.org/10.1016/j.mex.2023.102149 (accessed on 11th August 2025).

Proactive.IT Appointments Ltd. (2023). How AI can improve employee mental health and well-being. (2023). Proactive.IT Appointments Ltd. https://www.linkedin.com/pulse/how-ai-can-improve-employee-mental-health-4kede/ (accessed on 11th August 2025).

Impas, R. T. (2018). The moderating effect of time management on the relationship between quality of work life and personal effectiveness of junior high school teachers in Davao Region. *International Journal of Management Excellence, 12*(1), 1713–1726. https://doi.org/10.17722/ijme.v12i1.1039 (accessed on 11th August 2025).

Larivière, B., Bowen, D., Andreassen, T. W., Kunz, W., Sirianni, N. J., Voss, C., Wünderlich, N. V., & De Keyser, A. (2017). Service encounter 2.0: An investigation into the roles of technology, employees and customers. *Journal of Business Research, 79*, 238–246. https://doi.org/10.1016/j.jbusres.2017.03.008 (accessed on 11th August 2025).

Leitão, J., Pereira, D., & Gonçalves, Â. (2019). Quality of work life and organizational performance: Workers' feelings of contributing, or not, to the organization's productivity. *International Journal of Environmental Research and Public Health, 16*(20), 3803. https://doi.org/10.3390/ijerph16203803 (accessed on 11th August 2025).

Lopes, S. L., Ferreira, A. I., & Prada, R. (2023). Correction to: The use of robots in the workplace: Conclusions from a health promoting intervention using social robots. *International Journal of Social Robotics, 15*(6), 907–907. https://doi.org/10.1007/s12369-023-01021-0 (accessed on 11th August 2025).

Majumder, S., & Biswas, D. (2021). Fundamentals, present and future perspectives of quality of work life. In *Proceeding of First Doctoral Symposium on Natural Computing Research: DSNCR 2020* (pp. 271–279). https://doi.org/10.1007/978-981-33-4073-2_26 (accessed on 11th August 2025).

Mubin, O., Stevens, C. J., Shahid, S., Mahmud, A. Al, & Dong, J. J. (2013). A review of the applicability of robots in education. *Technology for Education and Learning, 1*(1). https://doi.org/10.2316/Journal.209.2013.1.209-0015 (accessed on 11th August 2025).

Penley, J. A., Tomaka, J., & Wiebe, J. S. (2002). The association of coping to physical and psychological health outcomes: A meta-analytic review. *Journal of Behavioral Medicine, 25*(6), 551–603. https://doi.org/10.1023/A:1020641400589 (accessed on 11th August 2025).

Robo Advisory Market Size, Share & Trends Analysis Report by type (Pure Robo Advisors, Hybrid Robo Advisors), by provider, by service type, by end-use, by region, and segment forecasts, 2022–2030. (2022). https://www.marketresearch.com/

Grand-View-Research-v4060/Robo-Advisory-Size-Share-Trends-32169039/ (accessed on 11th August 2025).

Sahni, J. (2019). Role of quality of work life in determining employee engagement and organizational commitment in telecom industry. *International Journal for Quality Research, 13*(2), 285–300. https://doi.org/10.24874/IJQR13.02-03 (accessed on 11th August 2025).

Saleem, F., & Malik, M. I. (2023). Technostress, quality of work life, and job performance: A moderated mediation model. *Behavioral Sciences, 13*(12), 1014. https://doi.org/10.3390/bs13121014 (accessed on 11th August 2025).

Santana, M., & Cobo, M. J. (2020). What is the future of work? A science mapping analysis. *European Management Journal, 38*(6), 846–862. https://doi.org/10.1016/j.emj.2020.04.010 (accessed on 11th August 2025).

Shanmuganathan, M. (2020). Behavioral finance in an era of artificial intelligence: Longitudinal case study of robo-advisors in investment decisions. *Journal of Behavioral and Experimental Finance, 27*, 100297. https://doi.org/10.1016/j.jbef.2020.100297 (accessed on 11th August 2025).

Srivastava, S. K., Medina-Sánchez, M., Koch, B., & Schmidt, O. G. (2016). Medibots: Dual-action biogenic microdaggers for single-cell surgery and drug release. *Advanced Materials, 28*(5), 832–837. https://doi.org/10.1002/adma.201504327 (accessed on 11th August 2025).

Vrontis, D., Christofi, M., Pereira, V., Tarba, S., Makrides, A., & Trichina, E. (2022). Artificial intelligence, robotics, advanced technologies and human resource management: A systematic review. *The International Journal of Human Resource Management, 33*(6), 1237–1266. https://doi.org/10.1080/09585192.2020.1871398 (accessed on 11th August 2025).

Warhurst, C., Mathieu, C., & Dwyer, R. E. (2022). *The Oxford Handbook of Job Quality*. Oxford University Press.

Warr, P., Sinclair, R. R., & Tetrick, L. E. (2012). How to think about and measure psychological well-being. In *Research Methods in Occupational Health Psychology*.

Xu, G., Xue, M., & Zhao, J. (2023). The relationship of artificial intelligence opportunity perception and employee workplace well-being: A moderated mediation model. *International Journal of Environmental Research and Public Health, 20*(3), 1974. https://doi.org/10.3390/ijerph20031974 (accessed on 11th August 2025).

Xu, S., Stienmetz, J., & Ashton, M. (2020). How will service robots redefine leadership in hotel management? A Delphi approach. *International Journal of Contemporary Hospitality Management, 32*(6), 2217–2237. https://doi.org/10.1108/IJCHM-05-2019-0505 (accessed on 11th August 2025).

Xu, X., Lu, Y., Vogel-Heuser, B., & Wang, L. (2021). Industry 4.0 and Industry 5.0—Inception, conception and perception. *Journal of Manufacturing Systems, 61*, 530–535. https://doi.org/10.1016/j.jmsy.2021.10.006 (accessed on 11th August 2025).

Zavolokina, L., Dolata, M., & Schwabe, G. (2016). FinTech–What's in a name? https://www.scopus.com/inward/record.uri?eid=2-s2.0-85011313931&partnerID=40&md5=151a0787ac5465c74ac6853ad3cf9027 (accessed on 11th August 2025).

Zhang, L., Pentina, I., & Fan, Y. (2021). Who do you choose? Comparing perceptions of human vs Robo-advisor in the context of financial services. *Journal of Services*

Marketing, 35(5), 634–646. https://doi.org/10.1108/JSM-05-2020-0162 (accessed on 11th August 2025).

Zhou, W. (2017). A research of influence on HR management of robots used in a large scale. *Journal of Nanjing University (Philosophy, Humanities and Social Sciences), 54*, 23–34.

CHAPTER 3

The Impact of Gender Diversity in Leadership on Job Satisfaction Among Women Employees in Developing Economies

MUNISH GUPTA,[1,*] KSHITIZ JANGIR,[2] MANDEEP SINGH,[1] and NASIMA MH CARRIM[3]

[1]*Chandigarh University, Gharuan, Mohali, Punjab, India*

[2]*Manipal University Jaipur, Jaipur, Rajasthan, India*

[3]*University of Pretoria, Pretoria, Gauteng, South Africa*

*Corresponding author

ABSTRACT

The trajectory of Industry 5.0 delineates an evolutionary continuum where human–machine symbiosis transcends conventional mechanization and engenders a human-centric paradigm of value creation. Within this milieu, robo-advisors have emerged as transformative instruments, not merely confined to financial services but increasingly embedded in organizational ecosystems to reconfigure quality of work life (QWL) and employee well-being (EWB). This chapter critically interrogates the role of robo-advisory platforms as algorithmic collaborators capable of mitigating occupational fatigue, augmenting decision-making, and facilitating sustainable work–life integration. Drawing upon contemporary literature and case illustrations, the analysis elucidates how robo-advisors harness artificial intelligence, machine learning, and emotional AI to personalize interventions, predict burnout, and enhance psychosocial resilience.

Robo-Advisors and Artificial Intelligence in Human Resources Management: Revolutionizing HR. Sanjay Taneja, Swati Gupta, Girish Lakhera, Mohit Kukreti, and Ercan Özen (Eds.)

DOI: 10.1201/9781779641410-3

The discourse further foregrounds the dialectics of promise and peril: while these technologies advance inclusivity, efficiency, and autonomy, they also precipitate ontological anxieties relating to surveillance, data privacy, and the potential erosion of human agency. By synthesizing perspectives from human–robot interaction, HRM, and organizational psychology, the study posits robo-advisors as catalytic agents in the epistemic reconstitution of workplace well-being. Ultimately, the findings contend that robo-advisors, if ethically designed and inclusively implemented, can serve as pivotal enablers in orchestrating a balanced, future-ready workforce within Industry 5.0.

3.1 INTRODUCTION

The concept of gender equality is an optimistic practice in organizations and societies. It deals with the issue of relegation, exclusion, and the portrayal of female employees. Cho and Barak (2008) state that developing a comprehensive organizational climate that values individual differences and respects intergroup diversities is the primary goal of gender diversity policies. On the other hand, organizational policies and practices based on gender diversity are now recognized as vital for organizational effectiveness. Gyekye and Haybatollahi (2015) specify that gender diversity policies have an impact on organizational behavior. In the management literature, gender diversity has gained momentum in the last two decades (Mousa and Puhakka, 2019).

The Social Exchange theory developed by P. M. Blau (1964) serves as the theoretical framework for organizational behavior. It suggests that the best way to motivate female workers is to show them financial and social appreciation. Female employees were encouraged to excel in the competitive era following industrialization (Avolio and Gardner, 2005). Organizations' effectiveness is evaluated based on their approach to their workforce. Therefore, it has become crucial for organizations to identify factors contributing to employee turnover intention (Moynihan and Pandey, 2016). Despite higher turnover intentions among women, female voluntary turnover and its antecedents have received little attention from the scholarly community (Clark, 1997; Gupta et al., 2023). Griffeth et al. (2000) found that lower job satisfaction is a decisive variable in "intent to quit." Therefore, addressing female employees' job satisfaction and motivation is essential because they significantly impact the organization's performance. Gender diversity has a long-term and profound impact on female employees' perceptions, reducing their psychological capital, which is necessary for organizational success. Developing employees' skills through human resource management (HRM) initiatives is critical to transforming them into human capital (Pradhan et al., 2018).

This study analyzes how gender diversity in leadership affects employee satisfaction. It aims to create a model that can be used to analyze the effectiveness of gender diversity in leadership positions. The chapter is structured as follows: the literature review comes first, followed by the research methodology and conceptualization of the model. The following two sections measure female employee satisfaction and analyze the results. The final two sections are the findings, study implications, limitations, and future scope.

3.2 REVIEW OF LITERATURE

Going through a literature review examining the impact of gender diversity, especially in leadership positions, on organizational outcomes such as firm performance, corporate governance, and employee job satisfaction, several studies have found a positive relationship between women in leadership roles and firm financial performance (Canyon and He, 2017). Specifically, high-performing companies benefit more from having female leaders compared to low-performing firms. In terms of governance, Korenkiewicz and Maennig, (2022) identified that gender diversity on corporate boards improves perceived job satisfaction among employees. However, the mechanism behind this effect remains unclear. There may be an indirect influence whereby diverse leadership institutes policies and cultures that promote inclusion, work-life balance, and employee growth opportunities. Limited research directly assesses the impact of female representation in upper management on female employees' job satisfaction. Much of the literature centers on diversity programs and generic inclusive policies. However, the psychological need for a fulfilling work environment that engenders loyalty and engagement requires leadership that values justice, appreciation, and care for individual needs (Malhotra, 2020).

This study aims to address this gap by identifying the specific attributes of gender-diverse leadership that positively influence female employee job satisfaction. Factors such as management support for work-life balance, advocacy for women's career advancement opportunities, and commitment to equitable compensation practices will be examined. A model will be developed outlining the relationships between these leadership drivers and job satisfaction. Understanding these connections has important implications, especially for emerging economies seeking to create positive and productive workforces. Enhancing gender diversity in upper management and leadership may be vital in improving national-level economic growth and global competitiveness.

3.2.1 GENDER DIVERSITY

Female employees had a fixed rate of employee involvement in the mid-19th century, and their jobs were clustered in specific, lower-paying jobs involving usually female staff (Oppenheimer, 1973). Due to the initial hurdles in education or profession, female employees compared themselves primarily to a female peer group in their workplace rather than men (Taylor, 2002). The most visible distinction between the two genders has been the prevalence of a gap in wages, known as the gender wage gap, that can quickly impact female employees' perceptions of justice and equity (Card et al., 2012; Gupta et al., 2023). Many organizations worldwide have attempted to address gender disparity in the workplace over the years by creating policies and programs to bridge the gap. For example, Abbott, Honeywell, Pfizer, Visa, and other companies have created programs that help women grow their careers. Another company, Kimberly-Clark, has witnessed a 66% rise in women in senior positions (Gender Diversity in Asian Workplace | GMO Research, n.d.). This is attributable to the company's initiatives to provide growth opportunities, training, and flexible working hours to guarantee that the women who work for them can meet their professional and personal obligations.

3.2.2 JOB SATISFACTION

Job satisfaction has been thoroughly researched in various circumstances, but research on female employees has long been a dark alley. Several researchers have discovered the causative factors of job satisfaction in the workplace. Job satisfaction is a dependent variable that depends on organizational features such as leadership style and organizational policy (Rosenfeld, 1978). Porter and Steers (1973) depicted that satisfaction from a job is a reflex to aspirations of features such as compensation, work quality, authority, responsibility, prestige, and so on, which were inveterate by Clark (1997); Hao and Wang (2022); Tziner and Vardi (1984). They contend that when employees are dissatisfied with their assigned work, uncertain about their privileges, working in an insecure environment, and dealing with uncooperative colleagues, it can contribute to a sense of disconnection from the organization. Additionally, it's important for organizations to prioritize employee satisfaction as it directly impacts their performance and the overall costs of hiring and training new employees (Gupta et al., 2023a; Vlosky and Aguilar, 2009). In their insightful research, they concluded that both male and female employees share comparable predictors of job satisfaction. The authors validated that there is no disparity in job satisfaction triggers, confirming that female employees are just as satisfied as their male counterparts when they achieve the same predictors. It is clear that both genders aim to achieve identical predictors for job satisfaction.

3.2.3 LEADERSHIP PERCEPTIONS

Females are more interpersonal and democratic leaders by nature, whereas males are autocratic (Evans, 2010). Due to the variety of preferences in leadership styles, opinions about what is supreme may evolve. According to hegemonic masculinity, males are believed to have the attributes most linked with quality management performance, such as confidence, competitiveness, enthusiasm, and emotional resilience (Sheaffer et al., 2011). These perceptions may contribute to the deficiency of female leadership at some of an organization's top levels. If females are regarded to be somewhat more capable than males, it goes to reason that they are unlikely to be endorsed to the same extent as their counterparts.

3.2.4 GENDER DIVERSITY IN LEADERSHIP

The board of directors is accountable for delivering a framework and strategy for the company by setting the organization's strategic direction and developing a vision and mission roadmap. The board of directors' surveillance function protects the firm's principal (stockholders) by restricting the agent's (management's) self-interested decisions (Ellstrand et al., 2002). The board of directors has a significant indirect impact on executive policymaking. It is critical to consider the way and the form the board of directors may play a role most effectively in sound policymaking. According to Filley et al. (1976), a board with low gender diversity can easily manage traditional issues; however, unique and novel issues are well solved by a board with high gender diversity.

The differences influence gender diversity in personality traits between male and female leaders. An individual's personality is often thought of in terms of how much they exhibit certain traits at either extreme. According to Feingold (1994), two key models address the immediate factors contributing to gender differences. The biological model is the first one (Zuckerman, 1991) that highlights innate and biological bases underlying gender differences in personality. Second, the sociocultural model (Eagly and Wood, 2016) focuses on the role of social and cultural factors in the emergence of gender differences in personality. The biological measures such as systolic blood pressure reactions, heart rate, serum cortisol, growth hormone, prolactin, progesterone, glucose, and triglycerides assess the neuroticism and openness personality dimensions (Arnetz and Fjellner, 1986; Soliemanifar et al., 2018). Socio-cultural factors such as race and ethnicity, religion, gender, and social class significantly impact agreeableness, extraversion, and conscientiousness (Eck and Gebauer, 2022). All these personality traits jointly constitute the 'Big Five model' (Weisberg et al., 2011). According to Rahmani and Lavasani (2012), a significant difference in the big five personality factors was found between females and males while studying the specific traits. Gender differences in personality traits and the value of measuring personality at the aspect level are often characterized by which gender scores higher. One of the best ways to get at the psychological differences between males and females that underlie leadership styles is to apply the 'Big Five model' to the study of personality. The Big Five framework is prevalent in these investigations (Weisberg et al., 2011). The 'Big Five' are five major personality traits which are shown in Figure 3.1.

'Neuroticism' denotes the experience of unpleasant emotions as an outcome of perceived threats and punishment. Females are inclined to be more neurotic, pleasant, and extroverted than males. Anxiety, despair, rage, and emotional lability are examples of unpleasant or negative emotions.

"Big Five" Personality traits by Weisberg		
	Neuroticism	Experiencing negative emotions like anxiety, depression, anger, and emotional lability because of perceived threats and punishment
	Agreeableness	Tendency to cooperate, maintenance of social harmony and be concerned about other people
	Conscientiousness	Sociability, Assertiveness and Positive emotionality
	Extraversion	Self-efficacy, Orderliness, Dutifulness, Achievement-striving, Self-discipline and Cautiousness
	Openness	Imagination, Artistic interests, Emotionality, Adventurousness, Intellect, and Liberalism

FIGURE 3.1 "Big five" personality trait model.

Source: Adapted from Weisberg et al. (2011).

However, anger is the only aspect of neuroticism in which males outperform females. The ability to collaborate, maintain social harmony, and be sensitive about others is represented by 'Agreeableness.' Agreeableness characterizes the propensity to cooperate, upkeep social cohesion, and be concerned about peer groups. 'Extraversion' is associated with sociability, gregariousness, and a congenial emotional state. For extraversion, the inequalities between men and women are the lowest. No significant gender differences have been found in 'Conscientiousness' or openness.

Females have been viewed (Saint-Michel, 2018) as 'transformative leaders.' These leaders are recognized for respecting the personnel's ambitions, instilling trust, and stimulating enthusiasm and motivation. Transformational leaders are committed to employee advancement and success (Dvir et al., 2002). More often than not, women are viewed as more capable of leading transformative change than men (Saint-Michel, 2018). Leaders with transformational qualities are those who inspire their teams to achieve great things by valuing their input, boosting morale, and instilling trust (Saint-Michel, 2018). In conclusion, transformational leaders care deeply about their staff members' growth and want them to achieve their full potential.

3.2.5 FEMALE JOB SATISFACTION

Herzberg's (1968) two-factor job satisfaction hypothesis formed the basis for the conception of perceived satisfaction. Based on this theory, satisfaction and dissatisfaction are autonomous domains rather than extreme extremes. Employees respond to their job environment by distinguishing job context elements known as "hygienes" and job content aspects known as "motivators," which define two independent dimensions. Employee satisfaction will result if the motivation needs are fulfilled. Nevertheless, if employees' motivational needs are unmet, they will experience no satisfaction but dissatisfaction. Likewise, the lack of hygiene demands leads to employee unhappiness, but their fulfillment results in employee no-dissatisfaction. Although sociocultural and economic reforms have been implemented in most African and Asian countries over the last two decades, gender diversity management principles remain neglected (Mousa and Ayoubi, 2019).

The most visible contrast between the two genders is the prevalence of wage disparity, which influences female employees' perceptions of justice and equality (Card et al., 2012). Various countries may see wage disparities (Tam and Zeng, 2014). In a study conducted by Sowmya and Panchanatham (2011) with determinants of job satisfaction among females, the compensation and growth aspect of work characteristics was discovered to be one of the most influential causes influencing the job satisfaction of female employees. Equal compensation for male and female employees equaled their position in the company, imparting empowerment. Aside from balanced compensation, the second most important determinant of female employee job satisfaction is the organizational aspect, which includes an emotional tie and a sense of pride.

In most cases, females in organizations have fewer job opportunities than males because supervisors, for complicated and challenging tasks, favor male employees over female equivalents (Moynihan and Pandey, 2016). These missed opportunities limit female employees' accountability and responsibility, which increases the organization's risk (De Pater et al., 2010). On the other hand, male workers are assigned ability-defining jobs that enable them to advance faster up the managerial hierarchy (Blau and Devaro, 2007).

Research has identified the determinants of job satisfaction for female employees, and other studies have concluded the presence of gender diversity in these established job qualities. According to the ILO Global Wage Report 2016-17, Indian female employees have one of the highest percentage pay gaps of 30% (ILO).

3.3 RESEARCH METHODOLOGY

Various studies have identified factors that affect employee satisfaction. Most of the factors are related to the policies of the respective organization, organizational culture, and employee relations. After properly reviewing the literature, we forwarded a questionnaire to respondents in various organizations across India, South Africa, and Bangladesh to develop the model. We received 513 responses from female respondents, which were analyzed, and constructs or variables were determined by Exploratory Factor Analysis using the 'Measurement Model.'

3.4 CONCEPTUALIZATION OF THE MODEL

Males and females have varied perspectives and enculturation, and these distinctions must not be overlooked in the evolution of leaders. The behaviors that may assist males in succeeding and progressing are not the same as those of women seeking the same (Budworth and Mann, 2010). Female-perspective research is needed to analyze, understand, and develop leaders and sophisticated women for management roles. The employment landscape is evolving, and organizations must be prepared to manage gender diversity challenges. In this article, to study the effect of gender diversity leadership on job satisfaction among female employees, factors affecting job satisfaction are proposed to be identified by taking a sample of female respondents in emerging economies like India, Bangladesh, and South Africa. A model identifying the impact of 'Top-level leadership diversity' and job-related policies on female job satisfaction is proposed to be developed. For the sustainable growth of an organization, job satisfaction is vital, and there is a need to identify the factors that can lead to optimizing specific determinants of organizational performance. Females may opt for a "power sharing" management style (Bradshaw et al., 2016) and provide a democratic decision-making process.

After proper review and discussion, factors were classified under the following groups:

1. Training and Development or Personal Development Opportunities
2. Incentive policies
3. Work-life balance (Working hours)
4. Supervision
5. Mutual respect
6. Recognition

3.4.1 TRAINING AND DEVELOPMENT/ PERSONAL DEVELOPMENT OPPORTUNITIES

Organizations provide employees various training and development opportunities, sometimes through on-the-job training and other times through off-the-job training. The four tiers of training assessment procedures include reactions, learning, behavior, and results. The word "reactions" refers to how the participants received the training and development programs. The second level of assessment gauges participants' learning. The third level evaluates how much participants have transformed as an outcome of the training. The last level shows the impact of the training on employees' productivity. Heydari et al. (2019) identified how training programs affected workers' productivity at Maskan Bank in the province of Hormozgon. They concluded that increasing educational support has the most significant impact on banks' profitability. Sainsbury's Supermarket Ltd (Karim et al., 2012) investigated the importance of training and after-training assessment for employee efficacy. Semi-structured interviews and survey data were gathered.

Khan et al. (2022) investigated the performance of bank staff in Karachi, Pakistan. They investigated whether training programs had a favorable impact on employee performance. They emphasized the importance of motivation, stress level, knowledge, and training style on employees in the banking domain in Karachi, Pakistan. Ikramina and Gustomo (2014) conducted almost the same type of research concerning training and development modules at PT Bank in Tabungan, Negara. Their goal was to use the Kirkpatrick model to identify difficulties in training and development programs. According to the findings, the most severe issue is level three: behavior. Their findings show that the individuals' behavior did not alter following instructions.

Aguta and Balcioglu (2015) explored the impact of training programs on the performance of the organization. The essential aspects influencing employee performance are working circumstances, supervisory style, and leadership. Effective training programs increase employee happiness while also improving organizational performance.

3.4.2 REMUNERATION POLICY OR INCENTIVE

Management seeks to promote the organization's strategic and interim goals by offering compensation for fulfilling those goals through differentiating pay schemes. Compensation comprises rewards and benefits that perform a significant role in the corporate sector and influence employee retention. An employee obtains compensation in exchange for valuable service after performing his or her duties as part of the employer-employee relationship. This comprises not only basic pay but also incentives such as bonuses, money for performing additional work, and incentives, as well as perks such as health benefits, pension plans, tuition fee reimbursement, recognition rewards, and in-house accommodation.

Compensation comprises intrinsic and extrinsic benefits, which can be monetary or non-monetary. Most studies have focused on only a few precise benefits, such as financial assistance, profit-sharing schemes, and work-family benefits, including flexible hours, part-time work, and reduced work periods (Mangel and Useem, 2000). However, further research into employee perception and need analysis is needed. Suggesting and defining the dimensions to assist and work on the acknowledged issues is critical.

The complete segment of compensation management further enlists various challenges. The common topics which are often included or discussed under the ambit of compensation management are compensation structure, level of pay, policy for raise in pay or increment policy (can also be included under recognition), other non-cash benefits or perks, equity in compensation, and other things like transparency in payment of organization related to compensation management, payment justice, supervisor involvement in the payment of compensation, etc.

3.4.3 WORK-LIFE BALANCE

Employees in all firms are gender diverse, with women holding almost equal responsibilities compared to the workforce makeup of the past. As a result of various studies focusing on this topic, great attention has been paid to women's work-life balance and job happiness. Among these studies is one that looked at the influence of expected work-life balance and job satisfaction on healthcare professionals and discovered that organizational engagement is critical for creating and enhancing organizational commitment among healthcare workers.

Work-life balance, job satisfaction, and organizational dedication were known to have a positive association. It is contended that institutional and custodial assistance are essential variables in male and female work-family conflict. It is said that work-life balance results from visiting friends and family, leisure, and inventive activities. Work-life balance can help to alleviate issues such as stress, anxiety, and worker illness. A balance between job responsibilities and spending time with friends and family is a fundamental component of a complete, healthy existence, and certain nations appear more conscious of this than others. It was shown that there is an inverse association between time demands at the organizational level and work satisfaction. It is argued that the government's employment and parental leave policies demonstrate that having fewer work hours and more comprehensive family leave provide a critical additional source of job satisfaction.

3.4.4 SUPERVISION

POS theory (Eisenberger et al., 2002), engrossed in Social Exchange Theory, contends that employees feel more motivated to do a good job when they believe their supervisors support them. They are likelier to go above and beyond when they feel their efforts are appreciated. This idea is supported by pragmatic research, which is linked to various outcomes, including employee well-being, extra-role behavior, higher organizational engagement, and employee retention. Knies et al. (2017) suggest two types of supervisor patronage: increasing workers' commitment through supportive behavior to advance employees' careers. These two primary areas in leadership literature emerge in supporting and developing leadership. The first emphasis point is to assist employees in their routine work, such as informing them about their well-being. The second emphasis point is to assist employees in their professional growth by holding consistent discussions about their training outlooks. We use these two focal elements as dimensions of supervisor assistance in our study.

POS theory and HRM literature (Purcell and Hutchinson, 2007) have all emphasized the relevance of perceived supervisor assistance and support in public leadership literature. According to these data, employee perceptions are vital to employee attitudes and performance. Further, there is a contrast between descriptive and evaluative perceptions. The former refers to employees' statements of supervisors' factual support, whereas the latter refers to an affectively charged evaluation of the supervisors' assistance.

3.4.5 MUTUAL RESPECT

Over the last decade, mutual respect and trust have emerged as important research topics in organizational behavior (Dirks and Ferrin, 2001; Rousseau, 1995). Respect has been identified in the diversity of professions (organizational science, psychology, and so on). Sociology, psychology, and so on) have emphasized the individual, group, corporate, and inter-firm levels. Further, they discovered the results in more favorable workplace attitudes (e.g., employee workplace satisfaction and dedication) and workplace behaviors (e.g., knowledge dissemination, organizational citizenship behavior, and performance results). For instance, individual contribution, team performance, and business outcomes are all important. Mutual respect and trust encourage cooperative behavior and network-based communication and organization, eliminate disputes, lower transaction costs, allow for the quick articulation of ad hoc work groups, and encourage effective crisis response. Respect culminates in trust. Viswanathan et al. (2004) has identified three significant aspects of trust: first, trust, the anticipation or confidence that is reflected in another person; second, trust necessitates a willingness to be sensitive and failing to meet expectations; and third, trust implies a degree of reliance, which indicates that the acts of others influence a person. As a result, workers will feel safer and more secure in their workplace interactions.

3.4.6 RECOGNITION

Recognizing and appreciating employees can enhance productivity and boost happiness. Research indicates that job satisfaction isn't just about financial rewards; non-monetary benefits like acknowledgment and intrinsic rewards are also essential (Nelson, 2005). Based on a survey of

200,000 workers, significant employee appreciation may enhance profitability, customer service stages, and staff engagement and satisfaction (Gostick and Sony Electronic Publishing Company., 2007). Acknowledging the efforts of employees results in better communication (workers are more likely to come up with solutions and new ideas), improved collaboration (employees are more likely to volunteer to help and go the "extra mile"), and decreased absenteeism and turnover (employees will experience greater job satisfaction and dedication) (Nelson, 2005). When employees feel recognized, they are more likely to share ideas, offer help willingly, and stay committed to their jobs. He concludes that increasing the frequency of contingent positive reinforcement daily improves quality and productivity. Positive reinforcement and acknowledgment from peers and management help employees. Recognition serves as a powerful motivator by fostering confidence, happiness, loyalty, and commitment among individuals, driving them to perform at their best. Acknowledging a job well done is a highly effective morale booster. Regularly recognizing and appreciating your staff is one of the simplest methods to maintain their satisfaction and productivity. The self-worth of many individuals is deeply intertwined with their work. Therefore, managers, company owners, and HR professionals should actively utilize recognition to elevate employee engagement.

Although employee motivation is greatly influenced by recognition, only about 40% of workers in North America feel appreciated for their excellent work or outstanding personal achievements. Managers frequently need to pay more attention to essential nonmonetary sources of compensation. Recognition may not be widely practiced, but a study (Nelson, 2005; Sharma et al., 2023) revealed that 78% of employees view it as essential to be acknowledged by their boss. Furthermore, 84% of managers believe that non-monetary recognition enhances performance as a reward. In addition, a striking 91% of managers believe that rewarding staff motivates them.

3.5 MEASURING FEMALE EMPLOYEE SATISFACTION

After considering the responses from the respondents, this study suggests the following model in Figure 3.2 to evaluate the impact of leadership diversity on female employee satisfaction.

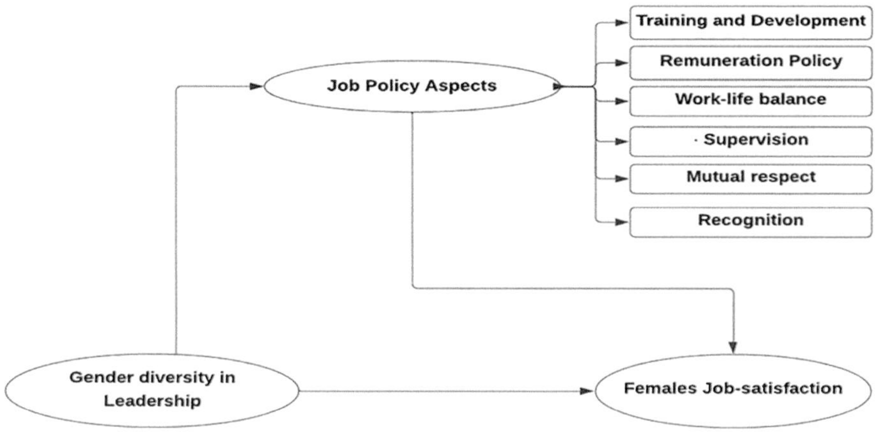

FIGURE 3.2 Model for measuring female employee satisfaction.

Source: Original Self Developed by authors.

Leadership diversity represents the diversity in leadership positions concerning gender, that is, the mix of male and female candidates in higher leadership positions. The leadership of an organization affects job policies, as well as the strategy formulation and policy formation of the organization. Leadership style and attitude affect the policies of an organization. Top-level management often carefully looks at human resources (HR), the primary resource of every organization. As per the earlier literature on the topic, gender diversity in the leadership of an organization affects its policies. Hence, leadership diversity affects an organization's employment policies.

Employment policies affect the satisfaction level of employees (Gupta et al., 2023b). It is the primary factor affecting employee satisfaction, attrition, and performance. Therefore, per the model, the policies will directly affect satisfaction. Leadership diversity can affect the satisfaction level of employees directly as well through employment policies.

3.5.1 *ANALYSIS AND RESULT*

The responses were evaluated in PLS-SEM version 4.1 to verify the model. The discriminant validity and reliability of the model were assessed. The values represent that validity and reliability are as per the limits analyzed in previous literature. Internal reliability under the measurement model, as well as discriminant and convergent validity criteria, was measured. Further, the HTMT Ratio (Henseler et al., 2015) and FL criterion (Fornell and Larcker, 1981) were followed to calculate the discriminant validity.

TABLE 3.1 Discriminant Validity Test

	HTMT values		
	Employee Satisfaction	**Job Policies**	**Leadership Diversity**
Employee Satisfaction	–	–	–
Job Policies	0.669	–	–
Leadership Diversity	0.702	0.599	–
	Fornell Larcker Criterion		
	Employee Satisfaction	**Job Policies**	**Leadership Diversity**
Employee Satisfaction	0.885	–	–
Job Policies	0.608	0.775	–
Leadership Diversity	0.596	0.495	0.769

Source: Self-computed by author.

TABLE 3.2 Reliability Test

	Cronbach's Alpha (CA)	C.R. (rho_a)	(AVE)
Employee Satisfaction	0.862	0.881	0.783
Job Policies	0.868	0.886	0.600
Leadership Diversity	0.766	0.800	0.591

Source: Self-computed by author.

The findings for discriminant validity are presented in Table 3.1. Here, the HTMT ratio falls below the threshold of 0.85, as suggested by Henseler et al. (2015). The Average Variance Extracted (AVE) values, identified in the table, exceed the correlation coefficients between the relevant factors. The reliability of the indicators for each construct has been measured using the Fornell-Larcker criterion, Cronbach's alpha, and Composite Reliability. Table 3.2 demonstrates that the Cronbach's Alpha and Composite Reliability value for each construct exceed the threshold limit of a minimum of 0.70 (Hair et al., 2017). Additionally, the convergent validity shown by the AVE scores for each construct exceed the benchmark of 0.50 (Henseler et al., 2009). The results suggest that the constructs explain more than 51% of the variance of the indicators (Franke and Sarstedt, 2019).

3.6 FINDINGS AND IMPLICATIONS OF THE STUDY

The study provides a comprehensive measurement model that explores the impact of gender diversity in leadership on the job satisfaction of female employees in developing countries. The model highlights a strong fit, indicating that these leadership factors are accurately identified and relevant in the context of job satisfaction of female employees. The model emphasizes the paramount importance of leadership prioritizing gender diversity, which decisively shapes the work environment and policies to influence the satisfaction levels of female employees positively. When job policies prioritize gender diversity as a critical element, they unequivocally serve as a bridge to enhance the satisfaction of female employees. This unequivocal mediation effect confirms that diversity and its implementation in job policies significantly impact employees' satisfaction (Behal and Gupta, 2022). The connection between having diverse gender leadership and implementing job policies that focus on the job satisfaction of female employees suggests that when companies in emerging economies embrace and put into action practices that promote diverse gender leadership, they notice a clear rise in the satisfaction levels of their female workforce.

Having women in leadership can be very positive for an organization. It can serve as an inspiration and representation for female employees at all levels of the organization. Studies have shown that women leaders tend to use leadership styles that are more inclusive, collaborative, and empowering. Employees value these behaviors because they lead to increased engagement and job satisfaction. When employees share similar experiences or backgrounds with their leaders, it often results in better relationships, higher satisfaction, and increased engagement. Organizations led by women are committed to prioritizing family-friendly policies, ensuring equal opportunities, and actively breaking down gender barriers. This inclusive corporate culture can significantly enhance employee satisfaction. A leadership team with diverse perspectives and thinking styles can enhance innovation and adaptability, while a culture that values innovation can lead to greater employee engagement and a clear vision. It's important to recognize that the relationship between gender diversity and employee satisfaction is complex and depends on organizational dynamics and leadership behaviors. Simply increasing gender diversity does not guarantee better employee satisfaction. It's important to create a fair and supportive environment in which female leaders and employees can thrive.

When organizations make an effort to value diversity, they can benefit from their employees' different perspectives and experiences, leading to overall success and satisfaction for everyone. The model proposes that it would be advantageous for emerging economies to take into consideration gender diversity as a vital factor in creating a supportive work environment. By fostering a more inclusive and rewarding workplace for female employees, organizations may have the opportunity to leverage the advantages of heightened diversity, ultimately contributing to enhanced decision-making, creativity, and overall productivity.

3.6.1 LIMITATIONS AND FUTURE SCOPE OF THE STUDY

The study is focused only on the specific factors identified in the chapter above when creating the research model. However, there might be other factors related to gender diversity that could significantly impact how satisfied female employees feel. The development of the model didn't consider important control variables like gender, age, educational background, and income level. Future research can look at these factors to understand how they could affect the outcomes. Adding these variables will help researchers better understand how well the model works for different groups of people.

The expanded approach holds potential to yield deeper insights and augment the model's predictive accuracy and relevance across diverse contexts. It is imperative to subject the model to cross-cultural testing within emerging economies to ascertain the universal applicability of its principles. Furthermore, the implementation of longitudinal studies is warranted to observe the enduring impact of gender-diverse leadership on employee satisfaction. Moreover, there is merit in extending the model to scrutinize the influence of combined control variables, including ethnicity, age, and educational background, on employee satisfaction.

3.7 CONCLUSION

Gender diversity in employment and decision-making (the focus of the current study) has lessened in recent decades. However, women account for less than a quarter of corporate board members in developing economies, although they make up nearly half, that is, 46% of the employed workforce. The proportion of females on the board may also impact the board's functioning. The impact of having more females on corporate boards on various performance measures has been extensively researched. This chapter explores the factors that can determine employee satisfaction for female employees. After exploring the factors, a model is suggested to determine how different factors affect employee satisfaction. The study collected data from organizations from more than 500 respondents spread across three other countries, that is, India, South Africa, and Bangladesh. Data was collected from only female employees as the study is only for the satisfaction of female employees. The developed model specifies that gender diversity in leadership positions affects the policies, which in turn can affect female employee satisfaction levels, and diversity in leadership is also directly related to satisfaction levels. The model was also verified, and discriminant validity and construct reliability were established. The developed measurement model provides insightful findings on the importance of gender diversity leadership and its impact on female employee satisfaction in emerging economies. It offers valuable implications for organizational policy-making and future research.

KEYWORDS

- **automation**
- **industry 5.0**
- **quality of work life**
- **robo-advisors**
- **well-being**

REFERENCES

Aguta, U. I., & Balcioglu, H. (2015). The impact of human resource management practices on organizational performance: A case of private banks in North Cyprus.

Arnetz, B. B., & Fjellner, B. (1986). Psychological predictors of neuroendocrine responses to mental stress. *Journal of Psychosomatic Research, 30*(3), 297–305. https://doi.org/10.1016/0022-3999(86)90006-1 (accessed on 11th August 2025).

Avolio, B. J., & Gardner, W. L. (2005). Authentic leadership development: Getting to the root of positive forms of leadership. *The Leadership Quarterly, 16*(3), 315–338. https://doi.org/10.1016/j.leaqua.2005.03.001 (accessed on 11th August 2025).

Behal, V., & Gupta, M. (2022). Reporting of corporate social responsibility practices: Evidence from Indian BSE-listed companies. *Prabandhan: Indian Journal of Management, 15*(3), 42–58.

Blau, F. D., & Devaro, J. (2007). New evidence on gender differences in promotion rates: An empirical analysis of a sample of new hires. *Industrial Relations: A Journal of Economy and Society, 46*(3), 511–550. https://doi.org/10.1111/j.1468-232X.2007.00479.x (accessed on 11th August 2025).

Blau, P. M. (1964). Justice in social exchange. *Sociological Inquiry, 34*(2), 193–206. https://doi.org/10.1111/j.1475-682x.1964.tb00583.x (accessed on 11th August 2025).

Bradshaw, P., Murray, V., & Wolpin, J. (1992). Do nonprofit boards make a difference? An exploration of the relationships among board structure, process, and effectiveness. *Nonprofit and Voluntary Sector Quarterly, 21*(3), 227–249. https://doi.org/10.1177/089976409202100304 (accessed on 11th August 2025).

Budworth, M. H., & Mann, S. L. (2010). Becoming a leader: The challenge of modesty for women. *Journal of Management Development, 29*(2), 177–186. https://doi.org/10.1108/02621711011019314 (accessed on 11th August 2025).

Card, D., Mas, A., Moretti, E., Saez, E., Autor, D., Dellavigna, S., Fisman, R., Hallock, K., Katz, L., & Oswald, A. (2012). Inequality at work: The effect of peer salaries on job satisfaction. *American Economic Review, 102*(6), 2981–3003. https://doi.org/10.1257/aer.102.6.2981 (accessed on 11th August 2025).

Cho, S., & Barak, M. E. M. (2008). Understanding of diversity and inclusion in a perceived homogeneous culture: A study of organizational commitment and job performance among Korean employees. *Administration in Social Work, 32*(4), 100–126. https://doi.org/10.1080/03643100802293865 (accessed on 11th August 2025).

Clark, A. E. (1997). Job satisfaction and gender: Why are women so happy at work? *Labor Economics, 4*(4), 341–372. https://doi.org/10.1016/S0927-5371(97)00010-9 (accessed on 11th August 2025).

Conyon, M. J., & He, L. (2017). Firm performance and boardroom gender diversity: A quantile regression approach. *Journal of Business Research, 79*, 198–211. https://doi.org/10.1016/j.jbusres.2017.02.006 (accessed on 11th August 2025).

De Pater, I. E., Van Vianen, A. E. M., & Bechtoldt, M. N. (2010). Gender differences in job challenge: A matter of task allocation. *Gender, Work & Organization, 17*(4), 433–453. https://doi.org/10.1111/j.1468-0432.2009.00477.x (accessed on 11th August 2025).

Dirks, K. T., & Ferrin, D. L. (2001). The role of trust in organizational settings. *Organization Science, 12*(4), 450–467. https://doi.org/10.1287/orsc.12.4.450.10640 (accessed on 11th August 2025).

Dvir, T., Eden, D., Avolio, B. J., & Shamir, B. (2002). Impact of transformational leadership on follower development and performance: A field experiment. *Academy of Management Journal, 45*(4), 735–744.

Eck, J., & Gebauer, J. E. (2022). A sociocultural norm perspective on Big Five prediction. *Journal of Personality and Social Psychology, 122*(3), 554–575. https://doi.org/10.1037/pspp0000387 (accessed on 11th August 2025).

Eisenberger, R., Stinglhamber, F., Vandenberghe, C., Sucharski, I. L., & Rhoades, L. (2002). Perceived supervisor support: Contributions to perceived organizational support and employee retention. *Journal of Applied Psychology, 87*(3), 565–573. https://doi.org/10.1037/0021-9010.87.3.565 (accessed on 11th August 2025).

Ellstrand, A. E., Tihanyi, L., & Johnson, J. L. (2002). Board structure and international political risk. *Academy of Management Journal, 45*(4), 769–777.

Evans, J. S. B. T. (2010). Intuition and reasoning: A dual-process perspective. *Psychological Inquiry, 21*(4), 313–326.

Feingold, A. (1994). Gender differences in personality: A meta-analysis. *Psychological Bulletin, 116*(3), 429–456. https://doi.org/10.1037/0033-2909.116.3.429 (accessed on 11th August 2025).

Filley, A. C., House, R. J., & Kerr, S. (1976). *Managerial Process and Organizational Behavior*. Scott, Foresman.

Franke, G., & Sarstedt, M. (2019). Heuristics versus statistics in discriminant validity testing: A comparison of four procedures. *Internet Research, 29*(3), 430–447. https://doi.org/10.1108/INTR-12-2017-0515 (accessed on 11th August 2025).

Gender diversity in Asian workplace | GMO Research. (n.d.). https://gmo-research.com/news-events/articles/gender-diversity-asian-workplace (accessed on 11th August 2025).

Gostick, A. R., & Sony Electronic Publishing Company. (2007). *The Carrot Principle: How the Best Managers Use Recognition to Engage Their People, Retain Talent, and Accelerate Performance*.

Griffeth, R. W., Hom, P. W., & Gaertner, S. (2000). A meta-analysis of antecedents and correlates of employee turnover: Update, moderator tests, and research implications for the next millennium. *Journal of Management, 26*(3), 463–488. https://doi.org/10.1177/014920630002600305 (accessed on 11th August 2025).

Gupta, M., Kumar, P., & Ghai, S. (2023). Impact of green finance on sustainability in India's commercial banks (pp. 216–233). https://doi.org/10.4018/979-8-3693-1388-6.ch015 (accessed on 11th August 2025).

Gupta, M., Sharma, V., & Jangir, K. (2023). An analysis of service quality measurement of the green payment service in India: A SERVQUAL model. In *Green Management–A New Paradigm in the World of Business*. Nova Science Publishers, Inc. https://www.scopus.com/inward/record.uri?eid=2-s2.0-85183524110&partnerID=40&md5=00cdd49879c85e455526dab3070a82b4 (accessed on 11th August 2025).

Gupta, M., Taneja, S., Sharma, V., Singh, A., Rupeika-Apoga, R., & Jangir, K. (2023a). Does previous experience with the Unified Payments Interface (UPI) affect the usage of central bank digital currency (CBDC)? *Journal of Risk and Financial Management, 16*(6), 286.

Gupta, M., Taneja, S., Sharma, V., Singh, A., Rupeika-Apoga, R., & Jangir, K. (2023b). Does previous experience with the Unified Payments Interface (UPI) affect the usage of central bank digital currency (CBDC)? *Journal of Risk and Financial Management, 16*(6). https://doi.org/10.3390/jrfm16060286 (accessed on 11th August 2025).

Gyekye, S. A., & Haybatollahi, M. (2015). Organizational citizenship behavior: An empirical investigation of the impact of age and job satisfaction on Ghanaian industrial workers. *International Journal of Organizational Analysis, 23*(2), 285–301. https://doi.org/10.1108/IJOA-08-2012-0586 (accessed on 11th August 2025).

Hair, J., Hollingsworth, C. L., Randolph, A. B., & Chong, A. Y. L. (2017). An updated and expanded assessment of PLS-SEM in information systems research. *Industrial Management & Data Systems, 117*(3), 442–458. https://doi.org/10.1108/IMDS-04-2016-0130 (accessed on 11th August 2025).

Hao, Y., & Wang, G. (2022). The effect of supportive organizational climate on employee turnover intention: A cross-level analysis. *Journal of Human Resource and Sustainability Studies, 10*(3), 334–355. https://doi.org/10.4236/jhrss.2022.103021 (accessed on 11th August 2025).

Henseler, J., Ringle, C. M., & Sarstedt, M. (2015). A new criterion for assessing discriminant validity in variance-based structural equation modeling. *Journal of the Academy of Marketing Science, 43*(1), 115–135. https://doi.org/10.1007/s11747-014-0403-8 (accessed on 11th August 2025).

Henseler, J., Ringle, C. M., & Sinkovics, R. R. (2009). The use of partial least squares path modeling in international marketing. *Advances in International Marketing, 20*, 277–319. https://doi.org/10.1108/S1474-7979(2009)0000020014 (accessed on 11th August 2025).

Heydari, M. R., Taghva, F., Amini, M., & Delavari, S. (2019). Using Kirkpatrick's model to measure the effect of a new teaching and learning methods workshop for health care staff. *BMC Research Notes, 12*(1), 1–5. https://doi.org/10.1186/s13104-019-4421-y (accessed on 11th August 2025).

Ikramina, F., & Gustomo, A. (2014). Analysis of training evaluation process using Kirkpatrick's training evaluation model at PT. Bank Tabungan Negara (Persero) Tbk. *Journal of Business and Management, 3*(1).

Karim, M. R., Nazmul Huda, K., Rehnuma, &, & Huda, K. N. (2012). Significance of training and post training evaluation for employee effectiveness: An empirical study on Sainsbury's Supermarket Ltd, UK. *International Journal of Business and Management, 7*(18), 141. https://doi.org/10.5539/ijbm.v7n18p141 (accessed on 11th August 2025).

Khan, M. Z., Shah, S. M., & Khan, A. A. (2022). Assessing conservation attitudes of mountain communities under different resource management regimes in northern Pakistan. *Environment, Development and Sustainability*, 1–21. https://doi.org/10.1007/s10668-022-02158-0 (accessed on 11th August 2025).

Knies, E., Leisink, P., & van de Schoot, R. (2017). People management: Developing and testing a measurement scale. *The International Journal of Human Resource Management, 31*(6), 705–737. https://doi.org/10.1080/09585192.2017.1375963 (accessed on 11th August 2025).

Korenkiewicz, D., & Maennig, W. (2022). Women on a corporate board of directors and consumer satisfaction. *Journal of the Knowledge Economy*. https://doi.org/10.1007/s13132-022-01012-y (accessed on 11th August 2025).

Malhotra, P. (2020). A relationship-centered and culturally informed approach to studying misinformation on COVID-19. *Social Media + Society, 6*(3), 2056305120948224. https://doi.org/10.1177/2056305120948224 (accessed on 11th August 2025).

Mangel, R., & Useem, M. (2000). The strategic role of gainsharing. *Journal of Labor Research, 21*(2), 327–343. https://doi.org/10.1007/s12122-000-1051-y (accessed on 11th August 2025).

Mousa, M., & Ayoubi, R. M. (2019). Talent management practices: Perceptions of academics in Egyptian public business schools. *Journal of Management Development, 38*(10), 833–846. https://doi.org/10.1108/JMD-01-2019-0030 (accessed on 11th August 2025).

Mousa, M., & Puhakka, V. (2019). Inspiring organizational commitment: Responsible leadership and organizational inclusion in the Egyptian health care sector. *Journal of Management Development, 38*(3), 208–224. https://doi.org/10.1108/JMD-11-2018-0338 (accessed on 11th August 2025).

Moynihan, D. P., & Pandey, S. K. (2007). Finding workable levers over work motivation. *Public Administration Review, 67*(4), 803–832. https://doi.org/10.1177/0095399707305546 (accessed on 11th August 2025).

Nelson, T. D. (2005). Ageism: Prejudice against our feared future self. *Journal of Social Issues, 61*(2), 207–221. https://doi.org/10.1111/j.1540-4560.2005.00402.x (accessed on 11th August 2025).

Oppenheimer, V. K. (1973). Demographic influence on female employment and the status of women. *American Journal of Sociology, 78*(4), 946–961.

Porter, L. W., & Steers, R. M. (1973). Organizational, work, and personal factors in employee turnover and absenteeism. *Psychological Bulletin, 80*(2), 151–176. https://doi.org/10.1037/h0034829 (accessed on 11th August 2025).

Pradhan, S., Jena, L. K., & Bhattacharyya, P. (2018). Transformational leadership and contextual performance: Role of integrity among Indian IT professionals. *International Journal of Productivity and Performance Management, 67*(2), 445–462. https://doi.org/10.1108/IJPPM-08-2016-0186 (accessed on 11th August 2025).

Purcell, J., & Hutchinson, S. (2007). Front-line managers as agents in the HRM–performance causal chain: Theory, analysis and evidence. *Human Resource Management Journal, 17*(1), 3–20. https://doi.org/10.1111/j.1748-8583.2007.00022.x (accessed on 11th August 2025).

Rahmani, S., & Lavasani, M. G. (2012). Gender differences in five factor model of personality and sensation seeking. *Procedia–Social and Behavioral Sciences, 46*, 2906–2911. https://doi.org/10.1016/j.sbspro.2012.05.587 (accessed on 11th August 2025).

Rosenfeld, R. A. (1978). Women's employment patterns and occupational achievements. *Social Science Research, 7*(1), 61–80. https://doi.org/10.1016/0049-089x(78)90004-2 (accessed on 11th August 2025).

Rousseau, D. (1995). *Psychological Contracts in Organizations: Understanding Written and Unwritten Agreements*. Sage Publications.

Saint-Michel, S. E. (2018). Leader gender stereotypes and transformational leadership: Does leader sex make the difference? *Management, 21*(3), 944–966. https://doi.org/10.3917/mana.213.0944 (accessed on 11th August 2025).

Sharma, V., Gupta, M., Jangir, K., Chopra, P., & Pathak, N. (2023). The impact of post-use consumer satisfaction on smart wearables repurchase intention in the context of AI-based healthcare information. In *Enhancing Customer Engagement Through Location-Based Marketing* (pp. 77–101). IGI Global.

Sheaffer, Z., Bogler, R., & Sarfaty, S. (2011). Leadership attributes, masculinity and risk taking as predictors of crisis proneness. *Gender in Management: An International Journal, 26*(3), 163–187. https://doi.org/10.1108/17542411111116563 (accessed on 11th August 2025).

Soliemanifar, O., Soleymanifar, A., & Afrisham, R. (2018). Relationship between personality and biological reactivity to stress: A review. *Psychiatry Investigation, 15*(12), 1100. https://doi.org/10.30773/pi.2018.10.14.2 (accessed on 11th August 2025).

Sowmya, K. R., & Panchanatham, N. (2011). Factors influencing job satisfaction of banking sector employees in Chennai, India.

Tam, V. W. Y., & Zeng, S. X. (2014). Employee job satisfaction in engineering firms. *Engineering, Construction and Architectural Management, 21*(4), 353–368. https://doi.org/10.1108/ECAM-10-2012-0098 (accessed on 11th August 2025).

Taylor, M. (2002). British politics in the age of revolution and reform, 1789–1867: Reform in Great Britain and Germany, 1750–1850. Edited by T. C. W. Blanning and Peter Wende. *Proceedings of the British Academy, 100*. Oxford University Press. Radicalism and revolution in Britain, 1775–1848. Edited by Michael T. Davis. Macmillan Press. Cornwall politics in the age of reform, 1790–1885. By Edwin Jaggard. *The Historical Journal, 45*(3), 661–677. https://doi.org/10.1017/s0018246x02002613 (accessed on 11th August 2025).

Tziner, A. E., & Vardi, Y. (1984). Work satisfaction and absenteeism among social workers: The role of altruistic values. *Work and Occupations, 11*(4), 461–470. https://doi.org/10.1177/0730888484011004005 (accessed on 11th August 2025).

Viswanathan, M., Ammerman, A., Eng, E., Garlehner, G., Lohr, K., Griffith, D., Rhodes, S., Samuel-Hodge, C., Maty, S., Lux, L., Webb, L., Sutton, S., Swinson, T., Jackman, A., & Whitener, L. (2004). Community-based participatory research: Assessing the evidence: Summary.

Vlosky, R., & Aguilar, F. (2009). A model of employee satisfaction: Gender differences in cooperative extension. *The Journal of Extension, 47*(2).

Weisberg, Y. J., De Young, C. G., & Hirsh, J. B. (2011). Gender differences in personality across the ten aspects of the Big Five. *Frontiers in Psychology, 2*, 178. https://doi.org/10.3389/fpsyg.2011.00178 (accessed on 11th August 2025).

Zuckerman, M. (1991). *Psychobiology of Personality*. https://books.google.com/books/about/Psychobiology_of_Personality.html?id=TA01Duy4RLwC (accessed on 11th August 2025).

CHAPTER 4

Empirical Study on Employee Engagement and Organizational Performance: HR Analytics Analysis

REVATI RAMRAO RAUTRAO[1,*] and NIKITA SURESH NILLE[2]

[1]*Dr D. Y. Patil B-School, Pune, Maharashtra, India*

[2]*G. H. Raisoni College of Engineering and Management, Pune, Maharashtra, India*

*Corresponding author

ABSTRACT

This empirical investigation interrogates the intricate nexus between employee engagement and organizational performance, foregrounding the pivotal role of HR analytics as a diagnostic and predictive instrument. Employee engagement, conceptualized as a multidimensional construct encompassing cognitive, emotional, and behavioral commitment, has been demonstrated to catalyze innovation, enhance productivity, and sustain organizational competitiveness. Drawing on primary survey data and secondary scholarly sources, the study scrutinizes how leadership effectiveness, recognition systems, work–life balance, and organizational culture collectively shape employee involvement. Through quantitative and qualitative analyses, the findings reveal a reciprocal relationship wherein heightened engagement fosters superior organizational outcomes, which in turn reinforce a culture of commitment and satisfaction. HR analytics emerges as a transformative enabler, furnishing actionable insights into performance indicators, absenteeism, individual and team

Robo-Advisors and Artificial Intelligence in Human Resources Management: Revolutionizing HR.
Sanjay Taneja, Swati Gupta, Girish Lakhera, Mohit Kukreti, and Ercan Özen (Eds.)

DOI: 10.1201/9781779641410-4

productivity, and overall workforce well-being. The integration of data-driven insights with human-centric strategies underscores the necessity of adopting adaptive HR practices to sustain long-term growth. Ultimately, this study posits that employee engagement, when systematically cultivated and analytically monitored, constitutes not merely a human resource prerogative but a strategic imperative essential to advancing resilience, agility, and sustainable organizational excellence.

4.1 INTRODUCTION

Team performance is essential when looking at employee engagement and organizational success from a wider angle. Employee engagement and, ultimately, organizational success are typically positively impacted by effective teams. Effective leadership is essential to fostering an inventive and imaginative culture. Leaders that embrace variety of thought, encourage taking risks, and provide instruments for testing foster an environment where people feel empowered to suggest new ideas. This leadership strategy not only increases employee participation but also sets up the framework for sustained organizational success via adaptability and ongoing improvement. This study is important because it has the potential to enhance how companies manage their workforces to achieve success.

Employees that are involved in the creative process feel purposeful and fulfilled. When employees operate in a positive, energetic environment where their ideas are valued, they'll be inclined to be pleased with their achievements and to be happier in their jobs. Encouragement for creativity is a powerful means of boosting staff morale because it gives people the chance to see the results of their creativity right away.

Employee engagement and satisfaction are correlated with organizational performance in a reciprocal manner. Prioritizing employee satisfaction as part of an organization's engagement strategy can lead to increased creativity, productivity, and teamwork. On the other hand, a positive workplace culture is reinforced by increased organizational performance, which is the result of happy and engaged workers, starting a virtuous circle of mutual gain.

4.1.1 INTRODUCTION TO EMPLOYEE ENGAGEMENT

An employee's expressive, intellectual, and motivational connection towards work with the organization they work for is embodied in the complex idea of employee engagement. It entails more than just being content with one's job; it includes feeling an intense sense of pride in one's work, being fully invested in organizational objectives, and being prepared to make an extra effort to help the business succeed. According to Larkin (2009), employee engagement is a contemporary business environment concern that is of great importance to both academics and business entrepreneurs.

Employee engagement is the idea that acknowledges workers as valuable assets whose happiness and well-being can have an important effect on organizational outcomes, instead of merely as passive contributors. Employee engagement, then, is essentially the result of a partnership that benefits the organization and its people alike by bringing together personal goals and organizational aims. A worker is considered engaged if they are passionate about what they do, share the organization's values and mission, and actively look for ways to improve both their own and the team's performance. Higher productivity, greater creativity, and a beneficial effect on the culture of the company as a whole are frequently the outcomes of this emotional investment.

Theoretically, employee engagement is based on a broad concept that explains why individuals are involved in their job and feel content. Employee involvement is all about establishing a work culture where each feels valued, connected, and passionate about what they do. According to the argument, engaged employees are seriously involved in their work and bring their best selves to work, rather than just going through the motions when performing their duties. This concept stems from the notion that work is about more than just tasks—it's about how people feel, how they relate to one other, and how much they believe in the company's objectives. The idea enables businesses to understand that when workers are motivated, they produce better work, and the business as a whole may prosper.

Businesses typically see greater profitability when their staffs are content, feel a sense of purpose in their work, and get along with one another. Employee engagement and the performance of the company are related in the same manner that when a team wins, it's due to everyone giving it their all. When employees are happy in their roles and collaborate well, the company can achieve enormous success! Employee engagement is crucial in the workplace, particularly for those in charge of an organization's human resources (HR). See HR as the helpful assistants that make sure everyone feels supported and content at work. Engaged workers enjoy what they do, have an attachment to the corporation, and collaborate well with others. For HR, this is like having a group of contented super-heroes on staff because motivated workers produce more, remain with the company longer, and contribute to a positive work environment. Thus, for HR professionals, ensuring employee engagement is equivalent to possessing a magic tool to create a phenomenal and prosperous workplace where all individuals may excel. It's similar to ensuring that everyone, every day, is happy and eager to work hard at the office!

4.2 OBJECTIVES

Research has attempted to investigate the relationship between organizational performance and the level of employee involvement, and also to analyze the aspects that affect employee involvement in an organization. Therefore, the objectives of the research are:

1. To determine different elements that affect worker engagement
2. To assess the current involvement of employees at stages.
3. To examine the impact of individual performance with involvement.
4. To assess efficacy of HR analytics in studying employee engagement.

4.3 REVIEW OF LITERATURE

Cattermole-Johnson (2014) state that workers who are motivated to contribute their expertise, abilities, and skills to the success of their company are called engaged employees. "The purpose of employee engagement is to ensure that workers are motivated to contribute to the success of the company, are devoted to its goals and values, and are able to improve their personal feeling of well-being."

Swarnalatha-Suresh Krishna (2013) The degree with which workers contemplate, sense, and behave generally demonstrates an increased degree of obligation towards the company is known as employee engagement.

Kruse (2012) states that an employee's expressive obligation to the company with objectives is referred to as employee engagement. An organization's productivity and profitability rise as a result of these initiatives to engage and encourage its workforce.

Bhatla (2011) argues that engagement can be linked to an emotional state and essentially speaks about a person's psychological commitment to the task at hand, which is evident in their work ethic.

Cook (2008) states that employee involvement is more of a mental agreement than a corporeal one. An engaged worker finds inspiration in their work. Consequently, it is upper management's duty to motivate and inspire staff members to do their jobs well.

Konrad (2006) states that employees' physical, cognitive, and emotional presentation of what they are in relation to their work and organization is a sign of engagement.

4.4 RESEARCH METHODOLOGY

A study is based on both the sources of data.

- **Primary Data:** Questionnaire.
- **Secondary Data:** Through books, published and unpublished theses, journals, magazines, periodicals.

4.5 NEED FOR THE STUDY

This study is significant because it clarifies the relationship between employee happiness and engagement and a company's performance. Employing intelligent technologies that examine employee data (HR analytics), we can determine what drives employee engagement and how this affects the performance of the business. Today's businesses want to be the best, and having happy workers who support their company's goals is one way to accomplish that. For this reason, having satisfied workers is quite important. This study can be used as a reference to help businesses increase employee satisfaction and enhance long-term business performance. It is about creating a positive work environment for all employees and ensuring business success as well. Setting specific objectives, utilizing tools for data analysis from surveys and other sources, and comparing the results with past practices is required for deciding if HR analytics is truly helpful in gauging employee satisfaction and engagement. We listen to workers' opinions about their experiences while continuously refining our usage of data. We can tell if something is effective if it contributes to employee satisfaction and business improvement. It is similar to applying artificial intelligence (AI) methods to determine employee satisfaction and its impact on business performance.

4.6 SIGNIFICANCE OF THE STUDY

This study is really important because it helps companies understand how happy and engaged workers make a big difference. Using smart tools that look at data from employees (HR analytics), we can figure out what makes employees happy and how this affects the company's success. This is a big deal because when employees are happy and do their best at work, the company does better too. The study is like a guide that can help companies make their employees happier, which in turn helps the company do well and be successful in the long run. It's all about making work a good place for everyone and making sure companies can be successful.

4.7 RESEARCH QUESTIONS

1. Describe the theoretical foundation for employee engagement.
2. In what way does organizational performance relate involvement of employees?

4.8 DATA ANALYSIS

4.8.1 DEPENDENT VARIABLES

4.8.1.1 ORGANIZATIONAL PERFORMANCE

It's critical to employ precise measures that support your research goals when measuring key performance indicators (KPIs) connected to organizational performance. Here are more thorough descriptions of the methods for measuring each KPI:

1. **Financial Performance (Table 4.1):**
   - Revenue Growth:

$$\text{Percentage Increase} = \frac{\text{Final Revenue} - \text{Initial Revenue}}{\text{Initial Revenue}} \times 100$$

TABLE 4.1 Financial Performance

Year	Revenue (Initial) (INR)	Revenue (Final) (INR)
2021	₹10,000,000	₹15,000,000

Source: Financial Data.

- Using the formula:

$$\text{Percentage Increase} = \frac{15{,}000{,}000\text{-}10{,}000{,}000}{10{,}000{,}000} \times 100$$

Percentage Increase = fifty percentage.

Hence revenue increased by 50% from 2021 to the end of the specific period.

TABLE 4.2 Employee Productivity, Quality of Work, Achievement of Goals and Targets, and Overall, Job Performance Ratings

Employee ID	Productivity (Units/Day)	Quality Score (Out of 100)	Goals Achieved (%)	Performance Rating (Out of 5)
001	20	90	95	4.2
002	18	85	80	3.8
003	22	92	100	4.5
004	17	88	75	3.5
005	21	94	85	4.0

Source: Survey Method.

- **Interpretation:** Examining the employees' performance indicators, it is clear that Employee 003 constantly exhibits high standards of

quality, productivity, and goal accomplishment; this is represented in the employee's excellent performance rating of 4.5. This person has a great work ethic and aligns with organization objectives, shown in their 100% goal attainment and 92 quality score. On the other hand, Employee 004 shows a lower level of productivity, having only achieved 75% of the goals established. This could suggest that further help or explanation regarding performance standards is required. Even though Employee 002's performance rating of 3.8 remains satisfactory, there is room for growth in terms of both productivity and quality. These observations underscore the significance of employing many performance measurements for an all-encompassing assessment, allowing establishments to pinpoint their strong points, tackle aspects that require enhancement, and customize training plans for every worker (Table 4.2).

TABLE 4.3 Profitability

Year	Full Income (INR)	Profit (Rs.)	Net Income (Rs.)
2021	₹20,000,000	₹5,000,000	₹3,500,000
2022	₹25,000,000	₹6,500,000	₹4,200,000
2022	₹25,000,000	₹6,500,000	₹4,200,000
2023	₹22,000,000	₹4,800,000	₹3,300,000
2024	₹28,000,000	₹7,200,000	₹5,000,000
2025	₹30,000,000	₹8,000,000	₹5,500,000

Source: Financial Data.

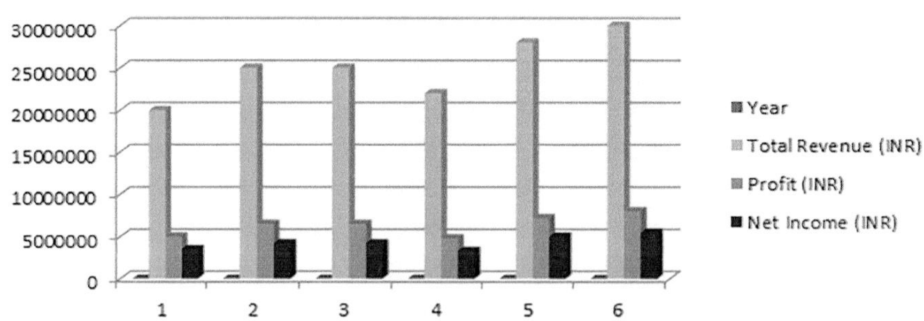

Significant trends may be seen in the financial performance over the five-year period. With ₹20,000,000 in total revenue in 2021, the organization made ₹5,000,000 in profit and ₹3,500,000 in net income. The following years saw significant expansion, with total income reaching ₹25,000,000 in 2022, ₹28,000,000 in 2024, and ₹30,000,000 in 2025. As profits and net incomes rise, so does revenue, which is expected to reach ₹8,000,000 and ₹5,500,000 in 2025, respectively. Remarkably, the company showed tenacity in 2023 by maintaining a respectable net income of ₹3,300,000 in spite of a decline in overall sales. Profits and net income have been steadily rising, which highlights the success of the company's financial strategies. However, the 2023 anomaly raises questions about how well the corporation manages costs during difficult times (Table 4.3).

4.8.1.2 LEADERSHIP

TABLE 4.4 Effectiveness, Work-Life Balance, Recognition and Rewards

Theme	Quotes	Insights
Leadership Effectiveness	"When our leaders are transparent and interact effectively our team truly flourishes. It significantly alters our level of motivation."	Positive leadership impacts team motivation.
Work-Life Balance	"I've been feeling overwhelmed lately. The deadlines are tight, and it's affecting my personal life. It's hard to stay engaged when you're constantly under pressure."	Poor work-life balance negatively affects engagement.
Recognition and Rewards	"Being recognized for your hard work is crucial. It boosts morale and makes you feel valued. It's a key factor for me in staying engaged."	Recognition is essential for employee morale and engagement.

The analysis results can be utilized to inform practical suggestions for enhancing employee engagement in your company (Table 4.4).

4.8.1.3 ORGANIZATIONAL STRATEGIES

TABLE 4.5 Themes, Representative Quotes, and Key Insights from the Survey

Theme	Representative Quote	Insights
Leadership	"Clear communication from leaders' enhancements team self-esteem."	Positive leadership is vital.
Work Environment	"A collaborative workspace enhances job satisfaction."	A positive environment is key.
Recognition	"Acknowledgment for hard work is unpredictable."	Improvements needed in recognition.
Communication	"Transparent communication fosters trust."	Open communication is valued.

Source: Survey Method.

The main ideas, illustrative quotations, and conclusions drawn from the qualitative data are compiled in Table 4.5. The analysis's conclusions can help develop organizational strategies that will increase worker engagement.

4.8.1.4 INDIVIDUAL PERFORMANCE

TABLE 4.6 Comprehensive Individual Performance Metrics

Employee ID	Productivity Score	Project Completion (%)	Customer Satisfaction (Scale: 1-5)	Innovation Contributions	Absenteeism Rate (%)
001	90	95	4.5	High	2
002	85	80	3.8	Moderate	5
003	92	100	4.7	High	1
004	88	85	4.2	Moderate	3
005	95	98	4.8	High	2
006	91	92	4.6	High	1
007	84	78	3.5	Low	6

Source: Survey Method.

- **Interpretation:** In terms of productivity, project completion, customer happiness, and innovation contributions, Employee 005 continuously outperforms all other KPIs. Excellent work is also displayed by Workers 001, 003, and 006, who range in their degrees of creativity and absenteeism. However, given the greater absence rate and decreased efficiency, project completion, and customer satisfaction scores, it is possible that Employee 007 needs additional help. HR and management may evaluate a wider range of workers and make well-informed decisions on recognition, growth, and support programs with the help of this expanded Table 4.6, which offers a more varied snapshot of individual performance.

4.8.1.5 TEAM PERFORMANCE

In an advantageous feedback loop, where engaged teams enhance organizational performance and organizational success in turn supports employee engagement, one can foster effective team dynamics, encourage working together, and align team goals with organizational objectives (Table 4.7).

TABLE 4.7 Comprehensive Team Performance Metrics

Team ID	Team Lead	Team Size	Project Success Rate (%)	Team Collaboration Score (Scale: 1-10)	Employee Satisfaction (Scale: 1-5)	Average Project Completion Time (Days)
A	101	8	92	8	4.5	30
B	102	10	88	7	4.2	40
C	103	12	95	9	4.8	25
D	104	6	84	6	3.5	50
E	105	15	91	8	4.7	35

Source: Survey Method.

With an impressive 95% project success record, a good teamwork score of 9, and excellent employee satisfaction (average rating of 4.8), Team C stands out as an exceptional performer. Additionally, Teams A and E score exceptionally well on a number of metrics, demonstrating strong teamwork and productive project outcomes. However, Team D has difficulties because of its lower project success rate, moderate collaboration score, and lower employee contentment. These findings imply that factors like staff satisfaction and collaboration are critical to attaining high project success rates. In order to increase team performance, organizations should give employee satisfaction and collaboration more weight because these factors can lead to better project outcomes and overall team success.

4.8.1.6 INNOVATION AND CREATIVITY

Creativity and innovation have a major role in influencing employee engagement and the success of an organization. Encouraging workers to think outside the box and submit fresh ideas boosts their personal engagement and has a big influence on the performance of the company as a whole. Developing a creative and inventive culture is one astute method of raising employee engagement and enhancing organizational performance. It is in line with the prevailing notion that highly motivated employees, given the freedom to innovate, may significantly boost a company's capacity for growth, adaptability, and overall success.

4.8.1.7 EMPLOYEE SATISFACTION

Employee happiness is a critical component that has a major impact on both organizational effectiveness and employee engagement. Contented employees foster a positive work environment that increases employee engagement and eventually boosts the overall performance of the company. Employee happiness is closely linked to a feeling of purpose, recognition, and a positive work environment—all of which are necessary for long-term commitment. Good customer service from happy employees positively impacts customer satisfaction and loyalty, two key elements that drive an organization's success.

Regularly asking workers about their feelings via brief questionnaires can contribute to a positive work environment. Use shrewd strategies to learn what makes individuals happy and improve their working environment. To show that you appreciate their opinions, pay attentively to what they have to say and take appropriate action. Give them little mementos to show your gratitude for all of their efforts. Encourage cooperation and respect for one another among the team members. Acknowledge that every individual has unique requirements and make an effort to be flexible when you can. You may avoid misunderstandings and enhance everyone's experience at work by speaking clearly. To create a welcoming and positive work environment, have an open mind and a readiness to learn.

4.8.1.8 CUSTOMER SATISFACTION

Keeping customers satisfied is crucial to our business. Periodically, possibly through surveys or comments, we should find out how satisfied they are. It's critical to trust their opinions and make any required adjustments in response to their critique. Customers who are happy with our services are more likely to come back and refer us to others. Thus, ensuring that our clients have a great experience dealing with us extends beyond just making sales.

4.8.1.9 ABSENTEEISM RATES

Regular absences from work are referred to as employee absenteeism. A thorough analysis of this is necessary for the team's recital. People who miss work on a regular basis could be discontent or might be dealing with other problems. We need to figure out what's going on and come up with better ways to support our team. Maybe it's about giving them more freedom or taking care of whatever concerns they might have. By keeping our absence rate low, we can make sure that everyone is doing well and that our job is going smoothly.

4.8.1.10 EMPLOYEE ENGAGEMENT

To promote employee engagement, make sure that everyone in the team is happy and excited about their work. Employees that are engaged genuinely care about the company and what they do. They're not only doing the tasks; they're putting in their best effort and getting along well with others. It feels like you got a bunch of super-enthusiastic superheroes on your side. To boost employee engagement, it's essential to pay attention to what they have to say, recognize their accomplishments, and create a positive work atmosphere. When everyone is involved, the team performs better, and the company grows.

4.8.1.11 QUALITY OF WORK LIFE

The feeling we get from our job is what defines the quality of our work life. Workplace happiness is more important than merely completing duties. A happy work life is one in which we enjoy going to work, feel valued, and have time for both personal and professional matters. It's similar to having a fulfilling job where we look forward to coming to work each day. We can ensure that everyone communicates effectively, exercise some flexibility, and foster a positive work environment to improve work life.

4.9 RECOMMENDATIONS

For the purpose of enhancing employee satisfaction and business performance, consider these straightforward HR analytics ideas: Utilize surveys to get input from staff members and utilize it to inform future improvements. Show appreciation for diligent work by implementing employee-friendly programs. Train and build skills in accordance with what analytics show workers require. Provide flexibility in your work schedule to help keep a work-life balance. Make sure leaders are productive by directing their growth with analytics. Analyzing collaboration patterns with analytics can help to promote teamwork. Implement physical and mental health programs to address the well-being of your workforce. Clear communication about objectives and modifications will foster trust. Use analytics to forecast attrition risks and develop retention strategies for important personnel.

4.10 SUGGESTIONS

Everyone finds the workplace to be more enjoyable when these things are done by all. Talk and listen to your co-workers, be amiable, support one another, take quick pauses to unwind, express gratitude for others when they do something kind, never stop learning new things, always be on time, share your ideas, remain upbeat even under pressure, and remember to take care of yourself outside of work. Conduct frequent surveys and feedback systems to learn about the opinions and worries of your staff. To get understanding from this information and make wise decisions, apply HR analytics. Create recognition initiatives to honor and recognize staff members for their accomplishments. HR analytics can be used to monitor and evaluate how these initiatives affect participation and output. To find areas that need work and talent gaps, apply analytics. Develop employee abilities through focused training and development initiatives, which will improve performance and engagement. Analyze the advantages and practicality of flexible work schedules using HR data. Acknowledge employee preferences and implement rules that support work-life balance.

To increase employee engagement and enhance company success, think about implementing these straightforward suggestions: Pay carefully to what your employees have to say, and make any required adjustments based on their suggestions. Say "thank you" to show your appreciation for all of their hard work. Make your office a friendly and positive environment to work so that everyone feels valued. Encourage cooperation and, when practical, exhibit flexibility. Use smart technology to find out what makes employees happy and improve their working environment. Increased team performance from a contented and driven staff results in company success.

4.11 CONCLUSION

In conclusion, businesses may raise employee engagement by using HR analytics and routinely conducting staff surveys to get input from employees. Analytics-driven recognition initiatives are effective in honoring and motivating employees. Acknowledging skill gaps and using analytics-driven targeted training will lead to improved performance. Additionally, if flexible effort plans are evaluated and put into practice utilizing HR data, they result in an effective balance. All things considered, integrating analytics into HR practices not only resolves current problems but also creates opportunities for ongoing improvements to employee engagement and organizational success.

Research on organizational performance and employee engagement has to include HR analytics in order to build a dynamic and flexible workplace. Regular surveys and feedback systems, when paired with HR analytics, offer valuable insights into employee attitudes and help make well-informed decisions. Analytics-driven recognition programs have shown to be effective tools for appreciating and motivating worker contributions. Customized training may be more easily provided with the use of analytics-based skill gap analysis, leading to increased performance. Moreover, employing HR analytics to assess and implement flexible work arrangements aligns with the evolving needs of employees for a comprehensive work-life balance. This comprehensive approach combines human insights with analytical technologies to help organizations continuously improve employee engagement, which raises organizational performance.

KEYWORDS

- **analytics**
- **engagement**
- **leadership**
- **performance**
- **satisfaction**

REFERENCES

AI Contentfy. (n.d.). The power of employee recognition for startup growth. https://aicontentfy.com/en/blog/power-of-employee-recognition-for-startup-growth (accessed on 11th August 2025).

Aswathappa, K. (2005). *Human and Personnel Management*. Tata McGraw-Hill.

Baldev, S. R., & Anupama, R. (2010). Determinants of employee engagement in a private sector organization: An exploratory study. *Advances in Management, 3*(10), 52–59.

Bhatla, N. (2011). To study the employee engagement practices and its effects on employee performance with special reference to ICICI and HDFC bank in Lucknow. *International Journal of Scientific & Engineering Research, 2*(8), 1–7.

Brown, S. C. (2017). Innovations in HR analytics. In *Proceedings of the International HR Conference* (pp. 45–56). HR Association.

Cattermole, G., & Johnson, J. (2014). Employee engagement is key in the hotel industry. http://www.hoteldesigns.net/industrynews/news_12787.html (accessed on 11th August 2025).

Chen, X., Rodriguez, M., & Patel, S. (2020). The impact of employee training on workplace innovation. *Journal of Innovation Management, 15*(4), 112–129.

Cook, S. (2008). *The Essential Guide to Employee Engagement: Better Business Performance*. Kogan Page Limited.

Doe, J. (2021). The impact of employee satisfaction on organizational performance. *Business Insights*. https://www.businessinsights.com/employee-satisfaction-impact (accessed on 11th August 2025).

Doe, J. A. (2017). Examining the relationship between employee engagement and organizational performance. *Journal of Organizational Psychology, 12*(3), 45–58.

Garcia, R. E., Clark, J. A., & Lee, S. Y. (2020). Analyzing the impact of training programs on employee engagement. *Journal of Training and Development, 22*(1), 89–104.

Gonzalez, L., White, E., & Turner, R. (2019). Analyzing employee satisfaction in the age of remote work. *Journal of Workplace Psychology, 22*(1), 45–62.

HR Analytics Institute. (2019). *Employee Engagement Trends Report*. HR Analytics Institute. https://hranalyticsinstitute.com/engagement-report (accessed on 11th August 2025).

HRAnalyticsPro. (2020). The power of HR analytics in employee engagement. *YouTube*. (accessed on 11th August 2025).

Johnson, M. B. (2020). Analyzing employee engagement through HR analytics. *Journal of Human Resources, 25*(3), 123–145.

Jones, S. D., Walker, T. R., & Davis, P. A. (2018). The impact of flexible work arrangements on employee well-being. *Journal of Occupational Health Psychology, 20*(4), 432–445.

Konrad, A. M. (2006). Engaging employees through high-involvement work practices. *Ivey Business Journal*.

Kruse, K. (2012). Why employee engagement? These 28 research studies prove the benefits. http://www.forbes.com/sites/kevinkruse/2012/09/04/why-employee-engagement/ (accessed on 11th August 2025).

Larkin, E. (2009). The challenge of employee engagement. http://www.hospitalitynet.org/news/4044076.html (accessed on 11th August 2025).

Lee, H. Y., Chen, J. A., & Kim, M. R. (2022). The role of leadership communication in fostering employee engagement. *Journal of Leadership Studies, 15*(2), 78–93.

Oliver, A. L., & Rothmann, S. (2007). Antecedents of work engagement in a multinational oil company. *SA Journal of Industrial Psychology, 33*(3), 49–56. https://doi.org/10.4102/sajip.v33i3.396 (accessed on 11th August 2025).

Robinson, D., Perryman, S., & Hayday, S. (2004). *The Drivers of Employee Engagement* (408). Institute for Employment Studies. http://www.wellbeing4business.co.uk/docs/Article%20-%20Engagement%20research.pdf (accessed on 11th August 2025).

Smith, J. A. (2018). *Employee Engagement and Organizational Success*. ABC Publishing.

Smith, J. A., Johnson, M. B., & Brown, S. C. (2020). Exploring the impact of HR analytics on employee engagement. *Journal of Human Resources, 28*(2), 75–89.

Swarnalatha, C., & Sureshkrishna, G. (2013). Role of employee engagement in building job satisfaction among employees of automotive industries in India. *International Journal of Human Resource Management and Research, 3*(1), 1–6.

University of Pennsylvania. (n.d.). Why communication is essential for effective leadership. https://lpsonline.sas.upenn.edu/features/why-communication-essential-effective-leadership (accessed on 11th August 2025).

Wang, L., Johnson, K. P., & Garcia, R. M. (2019). Impact of employee training on organizational performance: A longitudinal study. *Journal of Organizational Effectiveness: People and Performance, 24*(3), 210–225.

CHAPTER 5

Optimizing Work Performance and Employee Behavior by Using Artificial Intelligence and Machine Learning in India

ANAND SONI,[1] MOHAMMAD TAQI,[1] and RIZWAN KHAN[2,*]

[1]*School of Business, Bahrain Polytechnic, Isa Town, Bahrain*

[2]*University School of Business, Chandigarh University, Gharuan, Mohali, Punjab, India*

*Corresponding author

ABSTRACT

This study critically examines the transformative role of Human Resource Analytics (HR Analytics) in deciphering the dynamics of employee engagement and its consequential impact on organizational performance. Employee engagement, conceived as an amalgamation of cognitive alignment, emotional investment, and behavioral commitment, is increasingly recognized as a decisive factor in enhancing productivity, innovation, and corporate sustainability. Anchored in both primary data gathered through structured surveys and secondary sources from scholarly literature, the research delineates how variables such as leadership effectiveness, recognition, work–life balance, and organizational culture shape engagement levels. The analysis demonstrates a reciprocal relationship: engaged employees propel organizational outcomes, while organizational success further reinforces engagement. HR Analytics emerges as a critical enabler by offering data-driven insights into absenteeism, performance indices,

Robo-Advisors and Artificial Intelligence in Human Resources Management: Revolutionizing HR. Sanjay Taneja, Swati Gupta, Girish Lakhera, Mohit Kukreti, and Ercan Özen (Eds.)

DOI: 10.1201/9781779641410-5

productivity metrics, and workforce satisfaction. The findings accentuate the indispensability of integrating analytical precision with human-centric strategies, thereby fostering resilience, adaptability, and long-term value creation. Ultimately, the study posits that embedding HR Analytics into employee engagement frameworks not only enhances operational efficiency but also redefines engagement as a strategic lever for sustainable organizational excellence.

5.1 INTRODUCTION

In the modern corporate environment, companies use analytics to delve deeper into their data in order to increase profitability, gain a competitive edge, and raise productivity. Businesses are eager to leverage artificial intelligence (AI) and machine learning (ML) to get faster, more accurate results. AI and ML are changing an alarming number of areas so swiftly that, at the UN conference on sustainable development, one such advanced AI robot, Sophia, joined the panel and answered questions (Vilani, 2018). These days, hiring managers have a plethora of AI alternatives at their disposal, ranging from entry-level applications to sophisticated AI solutions. The study's conclusions provide some light on the evolution of AI and its implications for human resources (HR). Because of its influence on organizational performance and competitive advantage, AI has caught the interest of both firms and academia. This study set out to ascertain the association between employee engagement, a metric for the capacity of human, intangible, and material resources, and employee performance. This study demonstrates that employee engagement acts as a beneficial mediating factor in the relationship between AI and ML abilities and employee performance (Garima, et al., 2020). Because they are more committed to their professions, employees who work for companies with excellent AI capabilities are consequently more likely to provide high-quality work. It was discovered that more AI and ML capabilities are connected with improved worker performance. Lastly, the link between AI (direct or indirect) and employee engagement was considered a potential moderator for the indirect influence of AI skills on employee performance, but no significant association was found. The results of the survey indicate that professionals have a greater knowledge of the implications of AI. Managers ought to focus on the (good) experiences that workers have at work as a result of AI in addition to the technology's financial benefits.

Businesses may boost productivity and lower turnover by utilizing AI to raise employee engagement and general levels of positivity. It is important to promote awareness of this cultural potential.

AI and ML have become essential to corporate survival and growth in the modern business landscape. This paradigm change has streamlined business procedures across sectors and revolutionized HRM. AI and ML are used in HR for supply chain efficiency and talent development. This study uses secondary data from academic papers, publications, and surveys to investigate the expanding roles of AI and machine learning (ML) in HRM. The study examines the potential and emerging trends that AI and ML bring to human resource operations, while highlighting the growing importance of integrating these technologies into HR. In addition to being in style right now, AI and ML in HRM are essential to staying competitive in the ever-changing corporate world.

AI and ML are now transforming HRM. These technologies provide unparalleled HR process optimization, decision-making, and employee experience transformation (Garima, et al., 2020). This article examines AI and ML of the existing landscape, applications, and future developments in HRM as organizations realize their potential. This investigation demonstrates that AI and ML represent a paradigm change in HR operations rather than just technology advancements. Every facet of HRM is impacted by these technologies, including performance management and talent acquisition. By examining these technologies, we want to understand the opportunities and trends that will shape HRM in the future in a world where AI and ML predominate (Vilani, 2018).

For human resource management (HRM) to enhance organizational growth, it must adapt to new business areas. In order to better understand employee lifecycle management, general staff management, hiring and selection, onboarding, compensation management, and employee retention, this research examines the ways in which HR departments apply AI and ML (Kaplan et al., 2019).

5.2 NEED AND SCOPE OF THE STUDY

HRM, ML, and AI are transforming how businesses select, manage, and inspire employees. By using historical data and behavioral patterns, AI allows robots to make better judgments than humans. AI and ML technologies help businesses in every industry choose the best options to address problems more effectively. Therefore, when complicated problems are handled more deftly, the organization generates more. AI and ML technologies help firms make critical decisions and ensure their survival by preparing them for every eventuality. In the modern business environment, companies are employing analytics to delve further into their data to improve productivity, gain a competitive advantage, and bolster their bottom lines. Companies want quicker, more accurate results, which is why they are keen to combine AI with ML. AI and ML have the potential to be a vital tool for any company seeking quantitative support in decision-making, since they can analyze vast volumes of data, produce trend directions, and provide helpful recommendations. This research also offers insights into AI, which will usher in a new age in industry. Understanding the numerous elements, components, and functions of ML and AI in maximizing job performance and employee behavior inside a company is made possible by this study. This study also lists the advantages of AI and discusses the challenges in using ML and AI in HRM practices by business organizations.

5.3 ARTIFICIAL INTELLIGENCE, MACHINE LEARNING AND HUMAN RESOURCE MANAGEMENT (HRM)

In modern business, AI is crucial to HRM. As they fight to stay relevant, companies are turning to AI for resource upgrades. Computer-based intelligence is altering HR departments worldwide, from customized training programs to robotized recruitment systems that identify top talent. AI in human resources (HR) will improve work productivity significantly. Simulated intelligence will transform HR innovation (George et al., 2019). Companies across industries are using AI and ML. HR has advanced greatly because of simulated intelligence. HR professionals use software to manage hiring, onboarding, and training. Later, AI in HR will automate some of these cycles, allowing workers to deploy their energy and resources more wisely (Vivek et al., 2019).

AI is first used in recruiting. This cycle considers abilities, experience, capacities, and social fit to locate the best candidate for available positions. AI and ML can assist HR professionals identify these traits more accurately than resume screening or phone screening. Onboarding is where simulated intelligence can help. A competitor who is a good fit for an open position should be invited in and integrated with present employees. AI's true value is in enabling unique critique circles for each expert. Machine intelligence may assess an employee's strengths and weaknesses based on their own data (location, coworkers, job, etc.) and provide specific development suggestions. Machine intelligence allows managers to mentor staff more than ever (Aldulaimi, et al., 2020).

5.4 THE FUNCTION OF ARTIFICIAL INTELLIGENCE AND MACHINE LEARNING IN HRM PRACTICES

HR managers may save time and money by using AI to learn about "better employee experience" and time management. Since HR has a direct impact on employees' life, it is a crucial component of every business. Safe and well-organized work environments are essential for employees' success and utility. HR is responsible for making sure every worker feels safe, gets support, and has the opportunity, knowledge, and compassion to perform well. One of the newest technologies, AI, has enhanced HR. AI automates and completes the majority of low-value HR chores, freeing up time for the important job (Tewari et al., 2020).

5.5 RECRUITMENT AND SELECTION PROCESS

AI can improve HR recruiting and selection (R&S). Computer-based, intelligence-driven systems quickly identify qualified individuals. It can also ensure firms hire the best candidate by eliminating those who don't satisfy standards. Investing time and money ensures that the top competitors are hired for each case. Complex computations and AI skills allow AI to investigate competition information quickly and accurately. This lets it discover top competitors faster than spotters (Sharma, 2021). Man-made intelligence can use language analysis, mannerisms, and appearance to assess an applicant's work ethic and character (Pavlou, 2022).

Man-made intelligence can also assist in creating realistic expectations based on each position's skills and behavior. Man-made intelligence can also boost employer branding and reputation by promoting an organization's beliefs, lifestyle, and benefits across platforms. Man-made intelligence can attract competitors by recruiting more certified applicants, expanding the organization's deceivability and reach. Simulated intelligence automates and improves the screening processes by analyzing resumes, introducing letters, and portfolios using math and natural language processing (NLP). Simulated intelligence can improve employee experiences from ability management to enrollment by swiftly and accurately providing information (Dharmatti, 2022).

5.6 HOW DOES EMPLOYEE ENGAGEMENT THROUGH AI CAPABILITIES AFFECT EMPLOYEE PERFORMANCE?

This study employs an evaluative methodology to elucidate the influence of three AI capabilities—concrete, human, and abstract resources—on employee performance via employee engagement. There is a scarcity of study on the utilization of AI-driven technology for employee engagement. There is a lack of research that has investigated the relationship between AI, engagement among workers, and performance. This research extends beyond the realm of science, as it will provide guidance to managers regarding the capabilities of AI, worker engagement, and efficiency. It is imperative to enhance the value of AI and instruct leaders on how to inspire staff to embrace new technology.

5.7 STATEMENT OF THE PROBLEM

HRM is changing as a result of AI and ML. Businesses must have a thorough grasp of how new technologies affect the outputs and procedures of HR management when they implement them. The use of ML and AI in HRM has been the subject of several research studies. Nonetheless, in order to identify recurrent patterns, new trends, and areas requiring more research, a thorough examination of the available data is required. By analyzing many secondary data sources, this study attempts to address this gap by offering a comprehensive assessment of the potential, difficulties, and present status of AI and ML applications in HRM. The initiative's goal is to inform academics and business leaders on how new technologies affect workforce management.

5.8 REVIEW OF LITERATURE

The present study helps to explain how AI capabilities affect employee performance through engagement. Finally, a conceptual model with all methodology components is displayed. Dutta (2021) emphasizes AI's ability to transform employee experiences, whereas Davenport (2020) emphasizes HR's requirement for human intervention. Kshetri's (2021) Global South AI study shows it can improve recruiting and eliminate biases. Geetha and Bhanu (2018) examine how AI affects recruitment tactics, emphasizing human-AI collaboration. AI boosts HR activities, according to Yawalkar (2019). Martincevic and Kozina (2018) mention AI adoption obstacles and emphasize staff training. Gandhi et al. (2017) examined IT industry successes in integrating ML into HRM. Tambe et al. (2018) use Evidence-Based Management to solve AI implementation issues and provide guidelines. Jain (2017) examines AI's potential to change HR functions, and Charlier (2017) considers cost-effective talent acquisition.

Buzko et al. (2016) relate training to net income, promoting AI decision-making. According to Stone et al. (2015), Human Resource Information Systems (HRISs) will expand. Arntz et al. (2017) note that AI supports hiring and onboarding 24/7 and streamlines administrative tasks. Abraham et al. (2015) highlight AI's benefits to employers and candidates in the hiring process. Singh and Shaurya's (2021) mixed-method study in UAE companies shows that AI in HRM reduces routine tasks but also highlights training issues. Recent research shows that AI benefits HR. They concluded that AI is gradually taking over routine HR tasks with minimum human involvement. On the other hand, George and Thomas argued that AI is incapable of replacing human beings. Vivek and Yawalka also documented the ways in which AI decreases the amount of work required and enhances efficiency in the workplace. Jia, Guo, Li, and Chen utilized existing data to construct their reports, whereas George and Thomas conducted interviews with company HR personnel using standardized questionnaires. In their study of 115 HR experts and regional primary data, Garima, Vikram, and Vinay used Multiple Regression to evaluate the hypothesis.

These publications assert that AI is assuming control of numerous HR positions, although the authors (Garima, Vikram, and Vinay) and (George and Thomas) neglected to acknowledge the challenges faced by HR departments while employing AI tools. Jia, Guo, Li, and Chen expressed that the majority of businesses lack the necessary preparedness to adopt AI in their HR functions. On the other hand, Vivek and Yawalka highlighted the challenge of identifying suitable individuals to utilize AI tools, as well as the potential limitation of AI in HR decision-making due to the increasing dominance of technology. A thorough examination of incorporating this technology into the organization's HRM was not conducted. According to many authors, AI will have a positive impact on several activities within HRM.

Their reviews utilize secondary data, similar to ours, as we scrutinize articles, journals, blogs, and websites. The literature review identified deficiencies in the implementation of AI, career pathways, and prospective prospects, all of which will be addressed in this review paper. The concept of AI in the 1950s aimed to equip robots with cognitive abilities similar to those of humans (McCarthy et al., 2006; Pillai and Sivathanu, 2020). Gherhes (2018) believes advanced AI robots can provide new occupations and skills and solve societal problems by improving efficiency. ML is a branch of AI and uses statistical methods to learn from data. ML's advanced variant, Deep Learning (DL), uses a hierarchical technique to create complex data representations (Goodfellow et al., 2016). Recent ML advances are linked to DL (Lecun et al., 2015).

5.9 RESEARCH GAP

The field of AI and ML in HRM is expanding, but a comprehensive synthesis of findings from many secondary data sources is still lacking. The majority of research conducted today looks at particular issues, such as hiring and staff retention, rather than taking a more comprehensive approach. It is common for literature to be blind to the future, to fail to explain emerging patterns and developments. This research provides a comprehensive and up-to-date knowledge of the links between AI, ML, and HRM by examining available data from academic literature, case studies, and business reports.

5.10 OBJECTIVE OF THE STUDY

1. Examine global AI and ML applications in HRM.
2. Analyze trends, patterns, and optimal approaches for incorporating AI and ML into HRM processes, including hiring, educating, performance evaluation, and employee retention.
3. Analyze the challenges and constraints linked with the integration of AI and ML in HRM, encompassing ethical concerns, employee responses, and possible biases regarding decision-making algorithms, as evidenced in secondary sources.
4. Analyze secondary data to explore future patterns and implications of AI and ML on HRM, highlighting potential innovations, areas for improvement, and new technologies.

5. Critically evaluate the benefits, issues and challenges of AI and ML in HR management.

5.11 RESEARCH QUESTIONS

1. What are the areas of HRM where AI and ML tools may be used?
2. Why AI and ML is significant for the contemporary Human Resource Management?
3. What are the prospects for using AI and ML in the area of HRM?
4. What is the present status of implementing AI and ML tools in the field of HRM in India?

5.12 RESEARCH METHODOLOGY

This study employs descriptive research in this work. This work uses secondary data from academic papers, published materials, websites, and HR blogs. This study systematically synthesizes and analyzes secondary data on AI and ML integration in HRM. The narrative review format allows for literature, case studies, and reports to be examined. Starting with AI and ML fundamentals, the analysis moves on to HRM applications, and concludes with future trends and problems. A unified and comprehensive knowledge of HRM AI and ML evolution is provided by this chronological arrangement. This study used secondary data from different articles, magazines, journals, reports, case studies, and pertinent publications on AI and ML in HRM. To identify similarities and differences, data had to be categorized. Studies were compared to find gaps, discrepancies, and consistency in the literature. This method looked for in-depth understanding of AI and ML in HRM. In order to pinpoint areas in need of development and new technologies, anticipated advancements and innovations in AI and ML in HRM were examined.

5.13 ADVANTAGES OF IMPLEMENTING ARTIFICIAL INTELLIGENCE AND MACHINE LEARNING FOR EMPLOYEES

Decision-making is enhanced by AI/ML thanks to data-driven insights. Businesses may use AI and ML to quickly identify patterns and trends in vast volumes of data so they can make smarter decisions. By evaluating real-time inventory, order, and shipping data, an AI-powered supply chain system may enhance customer satisfaction, save costs, and optimize delivery timelines (Khatri, 2020).

AI/ML can give staff personalized insights and recommendations. Using an employee's work patterns, interests, and behavior, AI/ML algorithms can provide personalized training and development, job openings, and career routes. This improves job satisfaction and retention by helping people grow professionally (Melike et al., 2022). AI/ML can help doctors make better decisions. Medical AI can quickly analyze patient data, medical imaging, and other health data to deliver personalized therapy suggestions. Better patient outcomes, care quality, and healthcare costs can result (Kaur and Gandolfi, 2023).

AI/ML can automate repetitive tasks, improve decision-making, provide personalized insights and recommendations, and improve healthcare outcomes for employees. These technologies help employees operate more efficiently and effectively, improving job satisfaction, retention, and success.

5.14 ARTIFICIAL INTELLIGENCE CAPABILITIES

To maximize economic value from new technologies like AI, organizations must develop distinctive resources. 'A firm's AI competency is its ability to choose, orchestrate, and exploit AI-specific resources' (Mikalef and Gupta, 2021). As per the literature on information systems (IS), companies can achieve competitive advantages by effectively integrating and utilizing many firm-level resources that complement each other. These capabilities distinguish the firm from other organizational processes and are surpassed by competitors. Given the difficulty of identifying and establishing these attributes within the organization, rivals may find it challenging to replicate or diminish them (Morgan et al., 2009). In addition, they offer dynamic capacity, allowing organizations to actively modify their resource base. Dynamic capabilities refer to a company's capacity to effectively incorporate, generate, and adapt both internal and external competencies in order to respond promptly to rapidly evolving environments, as stated by Teece et al. (1997). Therefore, AI incorporates an organization's capabilities, information, processes, frameworks, and tactics. Therefore, AI readiness extends beyond technology (ALSheibani et al., 2018). This study investigates the AI capabilities of interpersonal skills, concrete resources, and abstract resources. It is important to acknowledge that these or other abilities may change in the rapidly evolving field of AI. The Figure 5.1 provides a comprehensive explanation of capabilities.

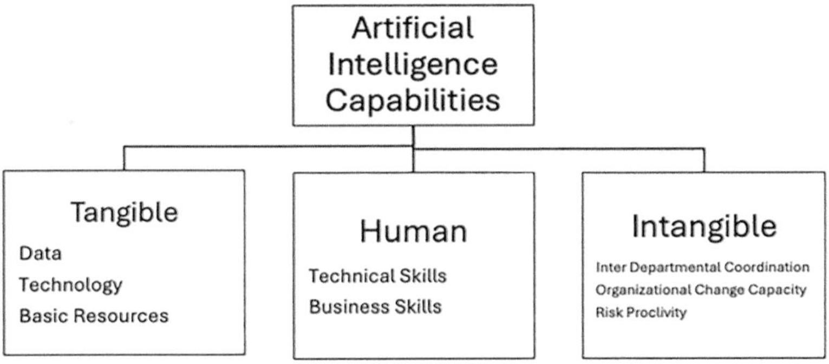

FIGURE 5.1 Artificial intelligence capability and categorization of resources.
Sourec: Adapted from Mikalef and Gupta, (2021).

5.14.1 TANGIBLE RESOURCES

Commercially accessible tangible resources include data, technology, and fundamental resources such as money and time (Mikalef and Gupta, 2021). Managers hold the belief that data enables them to optimize the potential of AI. For the development of AI algorithms, access to high-quality data is critical (Ransbotham et al., 2018). AI systems, similar to human beings, require vast quantities of superior training data in order to "learn" from the information at their disposal (Alonso, 2015; Bayless et al., 2020). These new data categories necessitate the development of novel technologies for data storage, processing, transportation, and protection throughout data collection, idea development, and application of AI training. Organizations must invest in systems capable of rapidly analyzing data and executing complex algorithms, in addition to flexible data storage (Dwivedi et al., 2021). AI technologies necessitate a number of infrastructure investments, which poses a challenge for many organizations, particularly those with limited funds. In addition to making financial and time investments in technological and data infrastructure, organizations must also have the means to support AI initiatives. As the majority of businesses are just beginning to practice with AI, it will require some time for these initiatives to mature and contribute impact (Ransbotham et al., 2018).

5.14.2 HUMAN RESOURCES

HR includes employees' knowledge, abilities or competencies, proficiency in experience, leadership, aims, interaction, collaboration, and problem-solving prowess. It includes business and technical abilities (Mikalef and Gupta, 2021). Technical AI skills are needed to create and implement AI algorithms, manage infrastructure, and introduce and ensure AI applications meet goals (Spector et al., 2019). Leaders must have the knowledge and willingness to drive a large-scale transition to maximize AI investments. Managers must also know AI's potential applications and how to migrate to AI-enabled operations (Fountaine et al., 2019). The competence of managers to begin and organize AI deployments is also important (Kolbjørnsrud et al., 2016). Managers must improve technical-line function collaboration to reduce friction and inertia. This will accelerate AI adoption and boost company value (Kiron, 2017).

5.14.3 INTANGIBLE RESOURCES

Many organizations struggle to replicate intangible resources, which are crucial in turbulent marketplaces. Intangible resources include risk proclivity, organizational change capacity, and interdepartmental cooperation (Mikalef and Gupta, 2021). According to Barney (1991), organizational abilities are subjective and socially intricate assets that are difficult to replicate, giving organizations a durable competitive edge. Current AI and business value study suggests that firms must foster cooperation, shared aims, and shared resources to maximize AI technology's value (Ransbotham et al., 2018). Fountaine et al. (2019) found that cross-functional teams with various talents generate AI with the best impact. By doing so, organizations can ensure AI programs serve corporate goals rather than specific business concerns. Furthermore, organizational change capacity refers to the potential issues that may arise if an old technique is not replaced. Creating a capacity to reduce obstacles to change and resistance is a key factor in transforming to digital capabilities and business value, as discussed in managerial research and research on IS (Besson and Rowe, 2012). An organization that cannot overcome this reluctance is unlikely to gain from AI investments. An organization with lots of data, skilled technicians, and cutting-edge AI infrastructure won't get performance gains unless it leverages these resources and changes its business model to incorporate AI (Mikalef and Gupta, 2021). Last, Fountaine et al. (2019) suggested that organizations should abandon their risk-averse approach and instead adopt a flexible, experimental, and adaptive mindset in order to reap the benefits of AI. The fundamental concept is that organizations that are inclined to depart from established protocols and establish ambitious objectives are more likely to develop robust AI capabilities compared to those who adopt a more cautious approach.

5.14.4 EMPLOYEE ENGAGEMENT

Employee engagement is a key management idea. To boost performance, companies are boosting employee engagement (Bailey et al., 2017). Employee engagement measures job satisfaction, value, cooperation, and trust. The statement asserts that employees that are actively involved and committed will remain longer and create better, more effective ways to contribute value. A successful company with happy employees and high production is the result (Catteeuw et al., 2007). Shrotryia and Dhanda (2020) define employee engagement as alignment, affectivity, and action orientation. According to Borges et al. (2021), literature reveals that strategic AI technology application for employee engagement is underutilized. They suggest investigating whether organizations may gain a competitive edge by using AI to engage employees. The concern is whether firms are overestimating the backlash to AI. The $10 billion spirits and wine giant Pernod Ricard found that AI complemented rather than replaced expertise and knowledge, so people embraced it. Companies that use technology well have more efficient and collaborative personnel. According to the Chief Global Strategy and AI Officer at Levi Strauss & Co., artificial intelligence (AI) has the ability to enhance employee engagement and foster creativity. A theory is being proposed:

- **H1:** The existence of capabilities of artificial intelligence, including tangible resources, human capital, and intangible resources, is directly correlated with higher levels of employee engagement.

5.14.5 EMPLOYEE PERFORMANCE

Employee performance is behavior that supports the company's objectives for labor quality, quantity, and timeliness. An organization's goods and services depend on work quality. Job quality means meeting requirements for product and service acquisition, manufacturing, quality inspection, and delivery. Job quantities include product, waste, and sales outputs from personnel. Job quantity (units of output) indicates an employee's commitment to their work; hence, monitoring it is important for work-related behavior. Finally, task time is the time needed to complete work-related tasks based on their difficulty. If tasks are done correctly and on time, and goods or services are provided, employees meet their working time goals. To conclude, employee performance affects an organization's outcomes, output, and success (Na-Nan et al., 2018). Companies can gain a competitive advantage in a world where humans and robots collaborate by implementing AI effectively throughout their business (Fountaine et al., 2019). A hypothesis is being proposed:

- **H2:** The performance of employees is favorably correlated with the capabilities of AI.

AI is said to boost employee performance. According to Cook (2008), organizations with engaged employees outperform their competitors. The literature links employee involvement to productivity, profitability, and more (Kumar and Pansari, 2014). Because they are engaged and invested in their work, passionate people perform better for companies (Reijseger et al., 2017). Employees who are enthusiastic about their work report increased profitability by 22%, customer satisfaction by 10%, and productivity by 21% (Ahmed et al., 2020). Employee engagement affects organizational performance; hence, it's vital in modern organizations. Several studies found that employee involvement improves job performance. Engaged employees are more likely to stay with their company, which may improve performance (Brenyah and Obuobisa-Darko, 2017). The following hypothesis is proposed:

- **H3:** There is a positive correlation among employee engagement and staff performance

The hypothesis asks if good employee performance is accomplished purely through AI capabilities or through employee involvement.

This mediating role of employee engagement has also been studied by Bailey et al. (2017) concerning work satisfaction and perceived organizational support, as well as person-organization fit and intention to leave, as discussed in Memon et al. (2014). Active involvement in their work increases the productivity of engaged employees (Reijseger et al., 2017). The following hypothesis is proposed:

- ➢ **H4:** Employee engagement acts as a mediator in the connection among the abilities of AI and employee performance

This hypothesis stems from the fact that AI encompasses a range of positions within organizations, including IT administrators, data researchers, software engineers, and systems analysts, as well as indirect roles like operational managers, managerial positions, and project managers. These various roles have the potential to impact employee engagement and performance.

5.15 SUMMARY AND DISCUSSION

Kshetri (2021) found that AI technology improves input efficiency, lowering R&S time and cost. AI improves production efficiency, especially by reducing application drop-off rates. This automation allows HR personnel to focus on strategic tasks, optimizing time management. Avature (2019) claims that because AI apps have an 80% application drop-off rate, they enhance HR procedures. According to Welbourne (2016), HR practitioners may improve the employee experience by using AI's ability to monitor and evaluate employees' levels of devotion and contentment. This might lead to improved rates of job satisfaction and retention. The findings of the study suggest that AI and ML might change recruiting, career path appraisal, problem detection, individualized corporate training, and employee engagement. It guides HR professionals through the process of technology integration by analyzing the benefits and drawbacks of adopting AI. The increasing usage of AI in HR and the evolving nature of human-machine collaboration are examined by Merlin et al. (2018). The study demonstrates how HR automation saves time and lessens workload. These studies come to the conclusion that businesses should use AI and that employees need to be upskilled in order to flourish (Table 5.1).

TABLE 5.1 Artificial Intelligence and Machine Learning Tools for Human Resources Professionals

Tool	Key Features	Price
Pymetrics	Job recommendations, gamified evaluation tasks, and candidate-job matching.	Undisclosed to the public pricing policy
Amazing Hiring	Collaborative recruitment, Templates for outreach communications such as email	Around $4800 per year, according to recent price trends.
ICMS Talent Cloud	Recruitment marketing via applicant tracking systems, Compatibility with additional HR platforms	Around $1700/month
Oracle Recruiting	Onboarding of personnel, analytics and reporting, and cloud implementation	Undisclosed to the public pricing policy
Skillate	Integrations with other ATS, candidate screening, insights and analytics	A non-restrictive and publicly undisclosed pricing policy
Textio	Real-time analysis of language, Real-time feedback, Optimization of job postings	Free to begin with, the fee becomes a consideration as usage continues.
Eightfold	Management of the candidate experience, support for career mobility, analytics and insights	Undisclosed to the public pricing policy
HireVue	Interviewer training, pre-employment evaluations, and video interviews	$35000 using a flat-rate structure
Entelo	Sourcing and engaging candidates, Prospective relationship management (CRM), diversity and inclusion	Pricing begins at $150 per month; no additional information is provided.
Talenture	Collaborative talent management and recruiting	The most base plan begins at $19 per employee per month.
iMocha	Coding interviews, candidate experience management, and analytics powered by AI	The price range may vary from $5000 to $25,000, contingent upon the specific undertaking.
Harver	Management of candidate experiences, scheduling video interviews, and pre-employment evaluations	Undisclosed to the public pricing policy
One Model	Customization, Data Visualization, and Employee Data Security	Undisclosed to the public pricing policy
Whatfix	Performance support, contextual assistance, and interactive guides	Policy regarding usage and project-specific pricing
intelliHR	Employee management, performance management, and compliance management	Four types of programs with monthly pricing varying by employee

Source: Vivek and Yawalkar (2019); Tiwari et al. (2021).

5.16 APPLICATION OF ARTIFICIAL INTELLIGENCE (AI) IN HUMAN RESOURCE MANAGEMENT (HRM)

AI helps to conduct comprehensive evaluations of many roles; it has the potential to modify HR operations. Virtual assistants are responsible for managing the entire process of recruiting, selecting, onboarding, managing performance, engaging employees, and ensuring their retention. AI applications rely on information systems for human resources (HRIS). HRIS is a system that gathers, saves, manages, retrieves, and verifies information on the human assets of an organization, employment activities, and characteristics of organizational units (Figure 5.2).

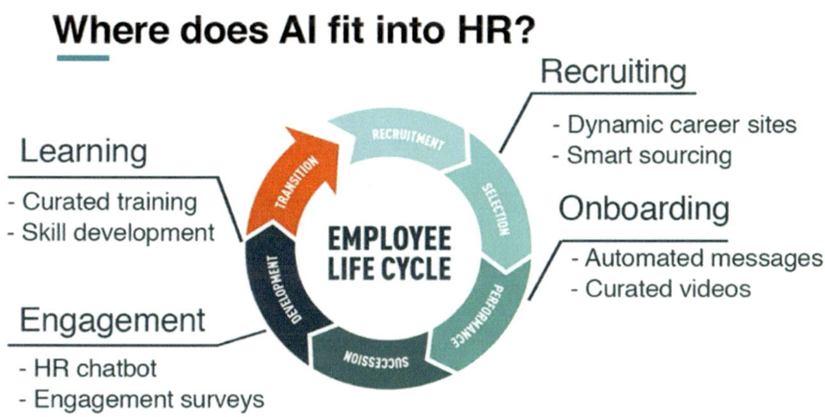

FIGURE 5.2 AI and HR.

Source: Compiled by Authors.

5.17 RECRUITMENT AND SELECTION

HR experts must hire the proper personnel for the company. Finding the perfect applicant among a pool of many skills is difficult. HR professionals may struggle to shortlist and evaluate resumes for prospective candidates. They must find the proper people to fill job openings quickly to avoid operational delays that could cost the company a lot of money. Giving candidates a nice experience boosts their chances of accepting the offer. Even as hiring needs rise, AI can speed up the process. It can automate monotonous operations by analyzing massive data sets for trends. It can also streamline hiring when recruiting. Websites can use chatbots to engage visitors and boost interaction. While interacting with bots, prospects will drop resumes and other basic info. Chatbots can inquire about candidates' desired roles and answer basic inquiries. Chatbots can collect candidate information, prequalify prospects, schedule meetings and chat times, and answer basic questions, saving time in recruiting (Mandeep et al., 2023).

ML can help evaluate large amounts of data and find trends the organization hadn't noticed. AI can screen resumes and find qualified candidates. ML-trained models examine candidates' experience, talents, education, and other company interests before hiring. This technology can identify applicants with the most appropriate talents. This will only examine qualified applicants and eliminate prejudices if designed appropriately. AI can also check candidates' social media profiles to guarantee they're qualified. This will save the recruiter time, maintain fairness, and hire the best candidate (Strohmeier et al., 2015).

5.18 THE PROCESS AND UTILITY OF HUMAN-AI CONTACT

Robotics enable the automated execution of tasks that involve strict regulations and frequent repetition. AI technologies, which mimic human collaboration, result in more complex recruitment procedures and more diverse recruitment situations. Advanced recruiting robots, like automated recruitment processes, resume screening algorithms, interview chatbots, and intelligent interview scoring systems, rely on AI technology to successfully accomplish the closed loop of human-AI interaction known as "perception-cognition-execution." AI can assist in developing effective strategies for company, optimizing operational scheduling, and ensuring quality control (QC). User input has uncovered implementation difficulties in order to enhance the interface between humans and AI (Abraham, 2015).

The robot assists individuals seeking employment and recruiters who desire flexible engagement guidelines within an inclusive environment. In order to meet customers' demands for intelligent robots, it is necessary to combine AI algorithms with other technologies such as Machine Sensing and Mechatronics, as there are multiple uncertainties involved in the interaction between humans and AI. Human-AI interaction can fully harness the capabilities of humans as a form of general intelligence to supplement AI, achieved through the mobility of robots and their active engagement with humans. The utilization of AI's interaction capabilities has the potential to enhance the quality of services and promote the integration of robots with intelligence in the arena of recruitment. The utilization of AI in recruitment processes is expected to increase as advancements in algorithm and hardware technologies occur (Chelliah, 2017).

5.19 THE EVOLUTION OF RECRUITMENT: FROM 1.0 TO 3.0 ERA

Job searchers used to explore job boards or newspapers until the mid-1990s. Once they discover a good position, they must visit the firm that posted it, get an application form for the job, fill it out manually, and submit it. However, job ads don't reach all interested job seekers, and recruiters are prone to cognitive biases. AI recruits faster in comparison to humans (Judge et al., 2000) (Figure 5.3).

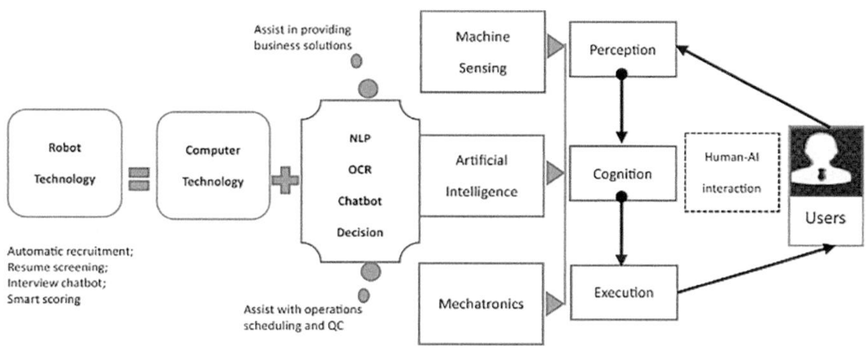

FIGURE 5.3 Human–AI interaction process and usefulness.

Source: Compiled by Authors.

1. **Digital Recruitment 1.0:** Internet digitization of work and candidate information expanded information extent and richness during the period from the middle to the end of the 1990s. Due to the absence of printing and delivery requirements, a comprehensive description of the position can be efficiently distributed to several candidates at a minimal expense. Job seekers are no longer required to manually search through physical job listings or send out several applications via mail.
2. **Digital recruitment 2.0:** It can collect job posts on many independent employment websites five years following the implementation of Digital Recruitment 1.0. Candidates do not need to browse every job advertisement, and recruiters can locate different job

seekers across all platforms without posting job information. Following the development of Digital Recruitment 2.0, Digital Recruitment 3.0 was introduced as a commercial product in 2015.

3. **Digital Recruitment 3.0:** AI has the ability to perceive speech, assess emotions, identify pictures, and make decisions by considering multiple elements (Ahmed and Reviews, 2018). AI algorithms have the capability to automate the process of recruitment. Algorithms and ML have the ability to rapidly examine data and identify patterns (Chichester and Gifen, 2019). The coming out of Digital Recruiting 3.0 can be attributed to background circumstances, including the substantial influx of applications per position throughout the 1.0 and 2.0 eras (Maurer and Liu, 2007).

As job application costs decreased and the quantity of applicants increased, firms had to spend more time assessing new aspirants. Talent's value is well recognized. Research is also showing that high-quality staff can give a company a competitive edge when tangible assets are its main asset (Paschen et al., 2020). This discrepancy shows that selecting the correct candidate from many applicants is crucial. In general, these smart "people" of the 3.0 Recruiting Era have changed recruiting.

AI aids organizations in identifying highly skilled candidates who possess both the necessary qualifications for the job (Person-Job Fit) and align with the values and goals of the organization (Person-Organization Fit) (Kristof-Brown, 2010). HRM derives advantages from its capacity to minimize and effectively eliminate obstacles in the processes of job acquisition and personnel financing. AI in HRM mitigates bias and nepotism during the procedure of applicant recruitment and selection. Throughout the process, a recruiter may be influenced by factors such as background, language proficiency, gender, and race. The integration between algorithm appraisal frameworks with automation and AI eradicates prejudices. An audit can mitigate or eliminate prejudice on this platform.

5.19.1 ONBOARDING

Onboarding swiftly and smoothly integrates new personnel into the company culture and policies. Onboarding is crucial to HRM. It defines and promotes organizational culture as well as demonstrating it. An effective onboarding process will make recruits like the company, stay involved, and want to stay longer. These recruits need greater attention, but individual attention is difficult. With the help of AI, the onboarding process may become self-service, which means that new hires can save time and effort by coordinating with the team and upper management without any human intervention. Early onboarding is crucial since recruits develop their first impression of the company.

5.19.2 TRAINING AND DEVELOPMENT

A good training facility is essential for a company's professional and technical staff. Create learning and development films with AI-based tools to automate learning. These training videos can be reused and translated without reshooting or rehiring voice actors. Video is one of the best ways to teach; thus, learners prefer it over text.

5.19.3 PERFORMANCE MANAGEMENT

A performance management structure is crucial for any organization. A strong performance management system tracks employees' job impact. With this system, the organization can track its training's impact. This framework will assist employees in matching job performance to organizational goals. Traditional performance management needs time-consuming stages, including creating goals, self-assessment, manager evaluation, discussion, and signing off. AI has the capability to continuously track objectives in real-time and offer feedback on completed tasks and outstanding ones. The evaluation approach can be integrated with performance management through the collection and analysis of employee performance data. AI solutions streamline the process of managing employee behavior and analyzing performance. AI has the ability to suggest incentives for achieving goals more quickly and send alarms and information on productivity when goals are not met.

5.19.4 COMPENSATION MANAGEMENT

This is essential to HRM. It involves analyzing, administering, and determining employee incentives and benefits. Company compensation and benefits affect employee retention. With today's fierce business competition, benefits and compensation are hard to stay up with. HRM must establish remuneration and benefits to satisfy organizational needs. Fair and competitive employee compensation helps organizations attract and retain top personnel.

5.19.5 EMPLOYEE RETENTION

In a competitive environment, losing a staff member lowers motivation and production. Employee turnover lowers revenue, raises recruitment and training costs, and delays growth. HRM must hire new employees to meet the staffing deficit. AI-powered solutions rooted in ML can help HR executives estimate staff attrition rates, identify leaving personnel, and provide rewards. Historical data is analyzed for patterns (Geetha et al., 2018).

5.19.6 CAREER PATH

HR training modules and learning management systems have long made use of AI to aid workers in charting a course for professional success, improving their performance in their present roles, and securing promotions. Thanks to this method, AI can make use of massive information like performance reviews, historical data, and biographies to build a better training and education model that is adapted to different levels of experience or occupation.

5.19.7 TRAINING AND DEVELOPMENT PROCESS

Computer-based intelligence helps companies systematically train their employees. Man-made intelligence can also offer workouts tailored to each person's needs for learning and understanding. Thanks to this, employees can keep up with the fast-changing corporate environment. AI is revolutionizing learning and development, helping specialists teach better and more engagingly. Man-made intelligence can help companies save time and money while improving employee learning. Intelligent learning and development solutions can give employees flexible advancement possibilities on demand (Jain, 2017).

5.19.8 EMPLOYEE APPRAISAL OR PERFORMANCE APPRAISAL PROCESS

This time of year, employees meet for performance surveys. Any organization needs a clear performance management system. Serious management strengths can assess employee influence on the workplace. This framework also shows the association's preparation's worth. This framework will benefit employees as they connect job performance to business goals. Traditional performance management involves tiresome steps including setting goals, self-assessment, manager evaluation, discussion, and closure. Manmade intelligence can help with objective observation and critique of what has and should be done in this situation. Computer-based intelligence increases employee performance correlation predispositions. Explicit, quantifiable goals can help. Man-made intelligence can also help set basic goals for employees who fail to fulfill performance expectations. It can also help determine employees' genuine potential and predict their performance. This data can help HR professionals plan advancement. HR professionals use simulated intelligence to set goals, evaluate performance, locate new data, and save functional time. This boosts efficiency and effectiveness.

5.19.9 EMPLOYEE ENGAGEMENT

AI can strengthen company-employee relations. Employee engagement projects like leading intelligent overviews, granting incentives and recognition, and providing constant input can use simulated intelligence. AI can boost employee engagement in these ways:

1. Simulated intelligence-powered tools can enhance employee experience by tailoring learning and development programs and HR strategies to individual needs and preferences. Employee satisfaction and engagement may increase.
2. Chatbots and remote helpers powered by simulated intelligence can provide continuous assistance to employees by answering questions and providing feedback. Creating a responsive and stable work environment can increase employee engagement.
3. HR professionals can use AI to analyze employee data such as engagement and performance. They can then use this data to identify concerns and familiarize themselves with the workforce. The results can be used to create employee engagement campaigns and programs that meet their needs and stress (Long Ho, 2022).

5.20 THE FUTURE PROSPECTS OF AI AND ML IN HUMAN RESOURCE MANAGEMENT

AI is expected to provide $13 trillion in worldwide economic activity by 2030, according to research. Richard Coombes, an HR transformation specialist at Deloitte, believes that AI could reduce behavioral and perceptual biases in human relationships. With the development of AI technology, HRM will become more automated, individualized, and data-driven. AI will improve the procedures of posting job openings, assessing candidates, and evaluating employee performance. AI will improve HR professionals' decision-making skills by providing current data and insightful analysis. In order to guarantee the moral and just use of AI algorithms, HRM should also plan for the emergence of new HR roles, such as AI ethics officers.

5.21 BENEFITS AND CHALLENGES OF AI AND ML IN HRM

AI reduces the time and effort needed to execute complicated activities, improving accuracy and results in many industries. More HR data means more data analysis time. AI software is now capable of identifying and managing complex data patterns, as well as performing critical tasks that require handling large amounts of data. This allows computers to detect issues and inconsistencies more quickly and accurately than HR. This saves time and boosts profits. Every company wants to make money; therefore, they maximize the benefits of AI and employee-machine collaboration (Jauhari, 2017).

AI also lets HR staff focus on more difficult tasks that require human intervention. This encompasses client interactions, workplace engagement, employee career development, and strategy. Many firms use online learning technologies for continual training; however, they are typically disorganized and don't assist employees. Using AI to arrange and present programs improves learning. AI can lower the cost of recruitment and training. It anticipates problems and offers useful solutions by analyzing and anticipating them (Long Ho, 2022).

AI will improve HRM in the future, but HR professionals must be mindful of the risks. Employee mindsets make AI integration into HR difficult. Individuals are increasingly concerned about AI's pervasiveness and ability to track employee behavior. Employees often struggle to understand and apply new AI tools due to their lack of technological abilities. As technology replaces HR's role in business decision-making, the HR department's capacity to make daily decisions may be limited (Pandey, 2019).

AI is devoid of emotional and psychological characteristics, which hinders its ability to keep track of human emotions and understand their impact on conduct, passion, and ambition. AI could be biased or inaccurate because it relies on human programming. Past data is essential for analysis and projections. AI "learns" from algorithms that analyze high-quality, bias-free data for analysis. The AI system acquires knowledge from the method and data it analyses and may inadvertently assimilate the biases of its programmer. Biased AI technology may produce biased outcomes (Pillai, 2020; Tambe, 2019).

5.22 PROSPECTS OF AI AND ML IN HRM

ML analytics is driving AI and ML in HR, which has bright prospects. This synergy helps HR professionals make informed decisions by revealing patterns and trends in large datasets. Predictive models can revolutionize several crucial areas. AI's predictive powers revolutionize recruitment by identifying ideal applicants for certain tasks. These models also help companies foresee employee turnover and execute retention tactics. Beyond recruiting and retention, AI-driven predictive analytics improve workforce planning, ensuring HR strategies meet changing organizational needs. According to Ho and Goethals (2022), AI in HR would enable strategic decision-making and organizational agility, not only automation.

Researchers expect AI will outperform humans in many areas in a few years. AI may surpass human competence and automate human work in the future years, say researchers. Some scholars believe AI will only benefit humans. People and machines will cooperate in the future. Another AI and DL study emphasizes the necessity of machine-human collaboration, where machines predict results and humans make judgments and act. If data is tracked, examined, managed, and protected, AI can play a significant role in HR. People analytics has plenty to teach and learn. As more companies adopt AI, competition for elite talent will intensify. Organizations will only differ in their capacity to meet the high digital expectations of candidates and deliver an exceptional experience.

The industry that trains its workforce to use AI and big data most effectively will emerge victorious (Yawalkar, 2019).

5.23 CONCLUSION

Recent business adoption has brought AI into HR. The details of our situation? Despite many wonderful improvements (driven especially by start-ups) and everything going digitally, HR seems to be limiting innovation. Modern HR pioneers employ AI for finance, recruitment, onboarding, and performance management. AI may eventually decide on hiring, firing, and other difficult tasks like answering HR questions and providing compensation statistics. Despite its benefits, computer-based intelligence in HR has several drawbacks, such as device limitations and cyberattacks.

Research predicts that AI would bring in $13.00 billion in global financial action by 2030. Manned intelligence is expected to reduce interpersonal behavior and perceptions, according to Deloitte HR change practice leader Richard Coombes. Long-term, computer-based intelligence innovation will increase computerization, customization, and information dynamic in HR management. Human resource practices including job posting, candidate screening, and performance evaluation will benefit from AI. By providing continual information and expertise, simulated intelligence can help HR professionals make better decisions. HRM should also prepare for new HR positions, such as AI morals officers who ensure that AI calculations are used ethically. As computer-based intelligence becomes incorporated into HR cycles, these professions will gain greater significance. HR experts must prioritize the fair and honest use of AI.

Importantly, the findings imply that while AI may threaten some white-collar tasks, it will enhance modern management's distinctly human characteristics. HR strategies require social and emotional intelligence, which are human domains. Cheng and Hackett (2021) suggest redefining HR Management with a data-driven approach and AI applications. As AI changes the HR landscape, future research should examine its wider impact on corporate performance and HR practitioners' changing roles. Training HR professionals in sophisticated numerical analysis and reasoning emphasizes strategic workforce preparedness. Finally, collective literature shows AI's HRM transformation potential. Organizations should intentionally integrate AI and modify their approaches to recognize the symbiotic relationship between technology and human skills. Adopting these technological advances is essential for HRM success in the dynamic and competitive company world.

KEYWORDS

- **analytics**
- **engagement**
- **leadership**
- **performance**
- **sustainability**

REFERENCES

Aggarwal, S., Mohamed Abdeldayem, S. M., & Aldulaimi, S. H. (2020). Trends and opportunities of artificial intelligence in human resource management. *International Journal of Scientific & Technology Research, 9*(1), 1. www.ijstr.org (accessed on 11th August 2025).

Ahmed. (2018). Artificial intelligence in HR. *International Journal of Research and Analytical Reviews*, IJRAR1944797. www.ijrar.org (accessed on 11th August 2025).

Aldulaimi, S. M., Mohamed Abdeldayem, S., & Hameed Aldulaimi, S. (2020). An empirical study of artificial intelligence and its impact on human resource functions. In *2020 International Conference on Computation, Automation and Knowledge Management (ICCAKM)*. Amity University.

Arntz, M., Gregory, T., & Zierahn, U. (2017). Revisiting the risk of automation. *Economics Letters, 159*, 157–160.

Artar, M., et al. (2022). Use of artificial intelligence in human resources processes. *7th International Zeugma Conference on Scientific Research.*

Artificial intelligence revolutionizes employee retention–GHD. (n.d.). https://www.ghd.com/en/perspectives/artificial-intelligence-revolutionises-employee-retention.aspx (accessed on 11th August 2025).

Avature. (2019). Does your candidate experience stand out from the crowd? www.avature.net/Infographics/CandidateExperience (accessed on 11th August 2025).

Bauer, T. N. (2010). SHRM foundation's effective practice guidelines series: Maximizing success. www.shrm.org/foundation (accessed on 11th August 2025).

Buzko, I., Dyachenko, Y., Petrova, M., Nenkov, N., & Tuleninova, D. (2016). Artificial intelligence technologies in human resource development. *Computer Modelling and New Technologies, 20*(2), 26–29.

Charlier, R., & Kloppenburg, S. (2017). Artificial intelligence in HR: A no-brainer. PwC. http://www.pwc.nl/nl/assets/documents/artificial-intelligence-in-hr-a-no-brainer.pdf (accessed on 11th August 2025).

Chelliah, J. (2017). Will artificial intelligence usurp white-collar jobs? *Human Resource Management International Digest, 25*(3), 1–3. https://doi.org/10.1108/HRMID-11-2016-0152 (accessed on 11th August 2025).

Cheng, M. M., & Hackett, R. D. (2021). A critical review of algorithms in HRM: Definition, theory, and practice. *Human Resource Management Review, 31*(1), 100698.

Davenport, T. H. (2019). Is HR the most analytics-driven function? *Harvard Business Review*. https://hbr.org/2019/04/is-hr-the-most-analytics-driven-function (accessed on 11th August 2025).

Dharmatti, M. (2020). Role of AI in performance management. *Medium*. https://medium.com/@manjunath.dharmatti/role-of-ai-in-performance-management-807e4fee5c77 (accessed on 11th August 2025).

Dharmatti. (2022). Artificial intelligence in HR or AI in HR. *Medium*. https://medium.com/@manjunath.dharmatti/artificial-intelligence-in-hr-3a11316e2ced (accessed on 11th August 2025).

Dutta, B. (2021). What is the role of artificial intelligence in human resource management? https://www.analyticssteps.com/blogs/what-role-ai-human-resource-management (accessed on 11th August 2025).

Garima, B., Vikram, S., & Vinay, K. (2020). An empirical study of artificial intelligence and its impact on human resource functions. *2020 International Conference on Computation, Automation and Knowledge Management (ICCAKM)*. Amity University.

Geetha, R., & Bhanu, S. R. D. (2018). Recruitment through artificial intelligence: A conceptual study. *International Journal of Mechanical Engineering and Technology, 9*(7), 63–70.

Gherheș, V. (2018). Artificial intelligence: Perception, expectations, hopes, and benefits. *Romanian Journal of Human-Computer Interaction, 11*(3), 219–230.

Goodfellow, I., Bengio, Y., Courville, A., & Bengio, Y. (2016). *Deep Learning*. MIT Press.

Jain, S. (2017). Artificial intelligence—The engine driving the next wave of transformation in business. *International Journal of Advanced Research in Science and Engineering, 6*(1), 592–596.

Jatobá, M., et al. (2019). Evolution of artificial intelligence research in human resources. *Procedia Computer Science, 164*, 137–142.

Jauhari, A. (2017). Artificial intelligence adaptation and its impact on human resources practices among the IT and ITES companies in Chennai. *Volume 7, Issue 9*.

Jiang, F., et al. (2019). Research on the application of artificial intelligence in human resource management. *2019 Annual Conference of the Society for Management and Economics, 4*, 258–262. The Academy of Engineering and Education.

Kaplan, A., & Haenlein, M. (2019). Siri, Siri, in my hand: Who's the fairest in the land? On the interpretations, illustrations, and implications of artificial intelligence. *Business Horizons, 62*(1), 15–25. https://doi.org/10.1016/j.bushor.2018.08.004 (accessed on 11th August 2025).

Kaur, M., & Gandolfi, F. (2023). Artificial intelligence in human resource management: Challenges and future research recommendations. *Review of International Comparative Management, 24*(3), 382–393.

Kaur, M., et al. (2023). Research on artificial intelligence in human resource management: Trends and prospects. *Global Journal of Management and Business Research, 23*(5).

Kovach, K. A., & Cathcart, C. E. (1999). Human resource information systems (HRIS): Providing business with rapid data access, information exchange and strategic advantage. *Public Personnel Management, 28*(2), 275–282. https://doi.org/10.1177/009102609902800208 (accessed on 11th August 2025).

Kshetri, N. (2021). Evolving uses of artificial intelligence in human resource management in emerging economies in the global south: Some preliminary evidence. *Management Research Review, 44*, 970–990.

LeCun, Y., Bengio, Y., & Hinton, G. (2015). Deep learning. *Nature, 521*, 436–444.

Long Ho, P., & Goethals, P. (2022). Machine learning applications in river research: Trends, opportunities, and challenges. *British Ecological Society, 13*(11). https://doi.org/10.1111/2041-210X.13992 (accessed on 11th August 2025).

Ma, L., & Ye, M. L. (2015). The role of electronic human resource management in contemporary human resource management. *Open Journal of Social Sciences, 3*(4), 71–78. https://doi.org/10.4236/jss.2015.34009 (accessed on 11th August 2025).

Martincevic, I., & Kozina, G. (2018). The impact of new technology adaptation in business. *Varazdin Development and Entrepreneurship Agency (VADEA)*.

McCarthy, J., Minsky, M. L., Rochester, N., & Shannon, C. E. (2006). A proposal for the Dartmouth summer research project on artificial intelligence, August 31, 1955. *AI Magazine, 27*(4), 12–14.

Merlin, P. R., & Jayam, R. (2018). Artificial intelligence in human resource management. *International Journal of Pure and Applied Mathematics, 119*(17), 1891–1895.

Pandey, S., & Khaskel, P. (2019). Application of AI in human resource management and Gen Y's reaction. *International Journal of Recent Technology and Engineering (IJRTE), 8*(4), 2277–3878. https://doi.org/10.35940/ijrte.D4585.118419 (accessed on 11th August 2025).

Pillai, R., & Sivathanu, B. (2020). Adoption of artificial intelligence (AI) for talent acquisition in IT/ITeS organizations. *Benchmarking: An International Journal, 27*(9), 2599–2629.

Singh, A., & Shaurya, A. (2021). Impact of artificial intelligence on HR practices in the UAE. *Humanities and Social Sciences Communications, 8*(1), 1–9.

Stanley, M. L., & Terry, L. J. (2012). An exploratory qualitative study of new employee organization onboarding from a human resource professional's perspective.

Stead, W. W. (2018). Clinical implications and challenges of artificial intelligence and deep learning. *JAMA, 320*(11), 1107. https://doi.org/10.1001/jama.2018.11029 (accessed on 11th August 2025).

Tewari, I., & Pant, M. (2020). Artificial intelligence reshaping human resource management: A review. In *2020 IEEE International Conference on Advent Trends in Multidisciplinary Research and Innovation (ICATMRI)* (pp. 1–4). https://doi.org/10.1109/ICATMRI51801.2020.9398420 (accessed on 11th August 2025).

Tiwari, P., et al. (2021). Application of artificial intelligence in human resource management practices. *International Conference on Cloud Computing, Data Science & Engineering.*

Vilani, C. (2018). What is artificial intelligence? Villani mission on artificial intelligence.

Vivek, M., & Yawalkar, V. (2019). A study of artificial intelligence and its role in human resource management. *International Journal of Research and Analytical Reviews, 6*(01), 20–24.

Wilkinson, T., & Redman, T. (2013). *Human resource management: A contemporary perspective.* https://www.researchgate.net/publication/320465545 (accessed on 11th August 2025).

Wilson, J., & Daugherty, P. R. (2018). How humans and AI are working together in 1,500 companies. *Harvard Business Review.* https://hbr.org/2018/07/collaborative-intelligence-humans-and-ai-are-joining-forces (accessed on 11th August 2025).

Wisetsri, W., et al. (2022). Artificial intelligence in human resources management: An overview. *Journal of Positive School Psychology, 6*(2).

CHAPTER 6

Artificial Intelligence: A Mediator in Recruitment Marketing

JYOTI THAKUR[1,*] and RUCHI SHARMA[2]

[1]*Center for Distance and Online Education, University School of Business, Chandigarh University, Gharuan, Mohali, Punjab, India*

[2]*Department of Management Studies, Graphic Era Deemed to be University, Dehradun, Uttarakhand, India*

*Corresponding author

ABSTRACT

Artificial Intelligence (AI) has emerged as a formidable catalyst in redefining recruitment marketing, reconfiguring conventional talent acquisition practices into data-driven, technologically mediated processes. By leveraging algorithms, predictive analytics, and intelligent automation, AI enhances candidate sourcing, resume screening, and applicant engagement, thereby accelerating decision-making and reducing transactional inefficiencies. The integration of AI-powered chatbots and virtual assistants fosters interactive, personalized candidate experiences, enabling dynamic pre-application communication and timely feedback. This research delineates the strategic significance of AI in strengthening employer branding and formulating proactive recruitment marketing strategies that extend beyond traditional post-application paradigms. Moreover, AI's capacity to anticipate hiring trends and identify high-potential candidates underscores its role as both an operational enabler and a strategic differentiator. While the findings illustrate profound benefits such as improved candidate-job matching, cost efficiency, and scalability, the study also acknowledges

Robo-Advisors and Artificial Intelligence in Human Resources Management: Revolutionizing HR.
Sanjay Taneja, Swati Gupta, Girish Lakhera, Mohit Kukreti, and Ercan Özen (Eds.)

DOI: 10.1201/9781779641410-6

attendant challenges, including ethical dilemmas, algorithmic opacity, and applicants' apprehensions regarding dehumanization in recruitment processes. By synthesizing insights from secondary data and scholarly discourse, the chapter posits AI not merely as a supplementary tool but as a transformative mediator in recruitment marketing, heralding a paradigm shift toward agile, inclusive, and technologically enhanced workforce acquisition.

6.1 INTRODUCTION

Artificial intelligence (AI) has revolutionized the field of recruitment marketing by integrating efficiency, data-driven decision-making, and better applicant experiences. Today, marketers can assess large datasets using AI algorithms in order to identify potential applicants on the basis of their skills, experience, and other relevant characteristics. Recruiters can identify qualified applicants faster and more efficiently due to the integration of AI into recruitment strategy. By data extraction and analysis of essential information such as skill sets, competencies, educational qualifications, and work experience, AI-powered applications can also automate the initial screening of resumes for further recruitment. Another technology, chatbots enabled by AI, has the capacity to interact with potential applicants on websites or their personal social media handles, addressing their queries and educating them about job specifications and requirements. AI-powered virtual assistants have the ability to collect preliminary information from candidates, schedule interviews, and provide a more dynamic and personalized experience to candidates as well as recruiters. AI can analyze past data to anticipate future hiring trends, enabling recruiters to make better data-driven judgments about where and how to allocate recruitment resources. Predictive analytics may also assist with successful employee recruitment, selection, onboarding, and retention projection. Today, AI has been acknowledged as a potent tool that holds the potential to create new opportunities for organizations, businesses, and services by opening new avenues for recruitment. This, in turn, is expected to contribute to the emergence of better, more efficient ways to recruit and retention methods, making AI a pivotal driver of innovation and development.

The intensifying competition for skilled employees has propelled AI's integration into the recruitment industry, not merely as an efficiency enhancer but also as a strategic advantage in order to acquire correct human capital (Bilal and Varallyai, 2019; Black and Van Esch, 2020). The capabilities of AI to process information and make decisions at a scale and speed surpassing human capacity further underscore its significance in enhancing job application likelihood. Integrating AI into recruitment advertising processes enhances accessibility for targeted candidates and acquires an apt workforce as per the job description, thereby augmenting competitiveness for organizations utilizing the approach (Buran Yi, 2018). Despite the targeted search and potential benefits of AI solutions, their widespread adoption in the recruitment field thus far is still limited. While there is a growing interest among organizations, the lack of knowledge, especially considering that many are new to AI solutions, impedes the swift entry of AI into the market (Bilal and Varallyai, 2019).

Although organizations utilizing AI in recruitment and their selection process are not yet the norm, the integration of AI tools is steadily gaining traction in the usage of AI to acquire access to the right aspirants. The escalating adoption of AI by managers places pressure on other organizations and their HR managers to accept and incorporate AI-based tools into their recruitment strategies in order to generate efficiencies. A study by Chuks Okolie, and Irabor (2017) reveals that experts in organizations utilizing AI software in recruitment experience benefits such as reduced costs, a higher number of applicants, improved candidate matching, streamlined application processes for candidates with diverse job opportunities, and an increased response rate in receiving feedback from applicants. Looking at the pace of AI adoption, the use of AI has the potential to not only metamorphose organizations internally but also impact the entire recruitment process and its way to reach aspirants, influencing the effectiveness of hiring a workforce.

6.2 RECRUITMENT MARKETING: STRATEGIC INNOVATOR

Recruitment stands out as one of the organizational functions poised for significant growth and change in the near future, as highlighted by Davenport and Ronanki (2018). Recruitment marketing involves a strategic and tactical approach aimed at enticing, involving, and cultivating potential talent before they formally submit job applications. Unlike the conventional reactive nature of recruiting, which initiates after a candidate applies, recruitment marketing is a proactive endeavor. The primary focus is crafting a compelling Employee Value Proposition (EVP) via creating a robust employer brand and getting suitable candidates to your organization. This proactive strategy is to create a positive perception of the company in the minds of the applicants, creating an environment that appeals to potential applicants even before the formal application process begins. Recruitment marketing extends beyond the traditional boundaries of recruitment and selection, extending its influence into the area of marketing and branding the brand to attract potential applicants with compatible competencies. It incorporates a strategic and tactical approach, leveraging various platforms to create a pre-application relationship with potential candidates for hiring. This chapter is divided into the multifaceted dimensions of recruitment marketing, analyzing its strategic significance, methods used, and the transformative impact that recruitment marketing has on traditional talent acquisition processes.

As businesses acknowledge the critical importance of attracting and retaining a top-tier workforce in an increasingly competitive global job market, the adoption of recruitment marketing becomes unavoidable. A data-driven recruitment marketing strategy not only fosters a positive image of the employer brand but also establishes the beginning of a relationship with potential candidates, increasing the likelihood of applicants joining the organization. This research chapter aims to explore the intricacies of recruitment marketing, dissecting its proactive nature, its role in helping employer brand creation in order to adopt new-age hiring technologies, and its contribution to the evolution of talent acquisition strategies. Through a comprehensive examination of industry practices, case studies, and academic literature, this research envisions highlighting the transformative power of recruitment marketing in shaping the contemporary dynamics of workforce acquisition.

6.3 METHODOLOGY OF STUDY

The chapter is descriptive in nature. The researcher gathered secondary information from numerous repositories, including newspapers, journals, theses, websites, case studies, reports, and magazines.

6.4 INTEGRATION OF AI IN RECRUITMENT MARKETING: THE EVOLUTION FOR GOOD OR NOT

In 2018, an evident trend among recruiters using AI software and solutions emerged in all business aspects, cementing AI's status as a key trend in the recruitment market as well (Esch et al., 2019). The ongoing boom in technical breakthroughs is continually redefining how organizations sell, use, and implement their online recruitment strategies, including job application and selection procedures, and AI is no exception. Job seekers are now inclined to align themselves with organizations that match their capabilities, offer competitive remuneration, provide effective training, and foster their personal growth graphs (Bilal and Varallyai, 2019). This introduces an additional layer to the application process, but candidates can analyze it by reviewing and applying directly through an organization's website portal recruitment tab or via online recruitment platforms like Indeed, Apna, Glassdoor, Shine, Seek, so forth. Whether hiring directly or through an intermediary, these platforms have the ability, and in some cases use AI, to filter, appraise, and match the most suitable individual with accessible employment opportunities.

While the incorporation of AI in online recruiting is a comparatively recent development in marketing methods, AI itself has been around for a while now. However, there is continuous debate over its use and definition (Faerber et al., 2003). The demand for outstanding performers is continually increasing. Finding the greatest resources is challenging, so it's no surprise that businesses are seeking artificial intelligence (AI)-based solutions to expedite their recruitment processes and gain a competitive advantage. AI in recruitment marketing can help optimize monotonous processes, discover suitable applicants, and provide crucial data to improve the overall recruitment strategy.

The overarching goal of AI is to automate and singularize repeated processes, identify qualified candidates, provide valuable insights, and autonomously conduct tasks independent of external sources, whether during or after the activity. The results of AI integration in recruitment marketing can be seen in the form of improved candidate screening rates, enhanced job matches, reduction in human biases, and data-driven decision-making. Companies like IBM, Unilever, and Intel have already integrated AI into their process of reaching out to the desired candidates. The utilization and promotion of AI represent a burgeoning trend in online recruitment marketing. While its current application primarily focuses on the initial reach to the targeted candidates and initial screening of applicant selection, the anxiety felt by job applicants regarding AI usage in recruitment is secondary to their perceptions toward the hiring organization.

There is limited information available thus far covering the candidate's response and feelings toward the incorporation of AI into the recruitment process and its marketing. According to Franklin and Graesser (1996), applicants tend to express greater satisfaction with technologically advanced features and procedures in online recruitment ads. Additionally, perceived efficiency toward the recruiter and user-friendliness of the ad platform, along with ad effectiveness, emerge as pivotal factors contributing to the overall satisfaction of the candidates with technology-based or AI-driven recruitment marketing processes (Irabor and Okolie, 2019; Sylva and Stefan, 2009). The amalgamation of candidate experiences with online recruitment technologies has the potential to substantially enhance AI recruiting technology and its worth. Establishing links between potential candidates, present employees, and organizations in an advanced recruitment environment may lead to an era of technology-mediated recruitment service networks that provide both constructive and immediate critique. Recruiters using AI-powered online recruitment technology can exploit this, allowing the system to identify the best-qualified applicant (Upadhyay and Khandelwal, 2018). This not only gives the candidate a timely indication, but it also benefits the business in terms of staff quality and ROI.

6.5 CONCLUSION

The widespread adoption of AI in recruitment marketing signifies an evolutionary change in how businesses attract and retain talent. The benefits of using AI in recruiting are clear, including increased efficiency, data-driven decision-making, and enhanced candidate experiences. AI algorithms enable organizations to scan through large datasets, discovering suitable applicants based on relevant characteristics, speeding initial screening through automated resume analysis, and interacting with AI-powered chatbots. The smart use of AI in recruitment advertising not only enhances accessibility for targeted candidates, but it also provides a competitive advantage in attracting the right talent. Despite growing interest, widespread use of AI in recruitment is hampered by a lack of understanding within organizations. However, early adopters' experiences demonstrate concrete benefits, including cost savings, higher applicant numbers, and more accurate candidate matching, simplified application processes, and increased responsiveness to feedback.

The research emphasizes the growing relevance of recruitment marketing, a proactive technique that extends beyond standard hiring procedures to build a favorable employer brand and nurture pre-application interactions with candidates. In the face of increasing competition for qualified workers, AI functions not just as a productivity increase but also as a strategic tool for recruiting the best human capital. Furthermore, the study digs into current trends in AI integration in the recruitment business, emphasizing its impact on job application and selection procedures. While the use of AI in online recruiting is a relatively new phenomenon, it has proven critical in optimizing processes, identifying quality candidates, and providing useful insights for overall recruitment strategy enhancement. Success Stories from IBM, so forth. Despite acknowledging the benefits of AI integration, the article emphasizes the scarcity of knowledge on applicants' reactions and attitudes towards this technology. Addressing candidates' issues and providing user-friendly AI-driven recruitment procedures are critical to maintaining overall satisfaction. The possibility for technology-mediated recruitment service networks, where AI enables constructive and quick feedback, is an exciting prospect for the future.

In essence, incorporating AI into recruitment marketing represents a positive advancement, providing organizations with a potent method for navigating the rapidly evolving waters of talent acquisition. As more organizations recognize AI's revolutionary potential, its widespread adoption is poised to revolutionize not only internal processes but the whole recruiting landscape, making it more dynamic, efficient, and sensitive to the demands of both employers and candidates.

6.6 FUTURE SCOPE OF STUDY

While AI misses the human empathy required to manage emotional situations, it is clearly capable of carrying out defined jobs and procedures more efficiently. By automating programmed recruitment tasks, AI incorporation can not only reduce the HR workload but also help to streamline processes and reduce the wastage of time and resources. Furthermore, it creates opportunities for researchers to explore more towards this emerging phenomenon. Researching into the development of newer recruitment marketing methods will assist researchers and businesses to attract top talent by highlighting their employer brand, culture, and company strengths. This can be accomplished through targeted advertising, social media integration into recruitment strategy, or workplace branding activities. Researchers and managers in the future can investigate more prospects of how AI may be used to personalize the recruiting experience for candidates, adapting job recommendations, messages, and interactions to their specific tastes and skills. Moreover, they can investigate the creation of predictive models utilizing AI and machine learning (ML) approaches to estimate talent requirements, identify high-potential applicants, and optimize recruitment efforts.

KEYWORDS

- **artificial intelligence**
- **automation**
- **employer branding**
- **marketing**
- **recruitment**

REFERENCES

Bilal, H., & Varallyai, L. (2019). Will artificial intelligence take over human resources recruitment and selection? *Network Intelligence Studies*, Romanian Foundation for Business Intelligence, Editorial Department, *13*, 21–30.

Black, J. S., & Van Esch, P. (2020). AI-enabled recruiting: What is it and how should a manager use it? *Business Horizons, 63*(2), 215–226. https://doi.org/10.1016/j.bushor.2019.12.001 (accessed on 11th August 2025).

Buran Yi, S. (2018). Dehumanizing, impenetrable, frustrating: The grim reality of job hunting in the age of AI. *The Guardian* [Online]. https://www.theguardian.com/inequality/2018/mar/04/dehumanising-impenetrable-frustrating-the-grim-reality-of-job-hunting-in-the-age-of-ai (accessed on 11th August 2025).

Chuks Okolie, U., & Irabor, I. E. (2017). E-recruitment: Practices, opportunities and challenges. *European Journal of Business and Management, 9*(11). www.iiste.org (accessed on 11th August 2025).

Davenport, T., & Ronanki, R. (2018). Artificial intelligence for the real world. *Harvard Business Review*.

Esch, V., et al. (2019). Marketing AI recruitment: The next phase in job application and selection. *Computers in Human Behavior, 90*(1), 215–222. https://doi.org/10.1016/j.chb.2018.09.009 (accessed on 11th August 2025).

Faerber, F., et al. (2003). An automated recommendation approach to selection in personnel recruitment. *AMCIS 2003 Proceedings*, 302. https://aisel.aisnet.org/amcis2003/302 (accessed on 11th August 2025).

Franklin, S., & Graesser, A. (1996). Is it an agent or just a program? In *Proceedings of the 3rd International Workshop on Agent Theories, Architectures, and Languages* (pp. 21–35).

Irabor, I. E., & Okolie, U. C. (2019). A review of employees' job satisfaction and its effect on their retention. *Annals of Spiru Haret University. Economic Series, 19*(2), 93–114. https://doi.org/10.26458/1924 (accessed on 11th August 2025).

Sylva, H., & Stefan, T. (2009). E-recruitment: A study into applicant perceptions of an online application system. *International Journal of Selection and Assessment, 17*(3), 311–323. https://doi.org/10.1111/j.1468-2389.2009.00473.x (accessed on 11th August 2025).

Upadhyay, A. K., & Khandelwal, K. (2018). Applying artificial intelligence: Implications for recruitment. *Strategic HR Review, 17*(5), 255–258. https://doi.org/10.1108/SHR-07-2018-0051 (accessed on 11th August 2025).

CHAPTER 7

A Conceptual Analysis on HR's Sustainability in the Age: Artificial Intelligence and Machine Learning

MUSKAN SINGH,[1,*] SACHIN GHAI,[1] AMAR KUMAR MISHRA,[2] and LUAN VARDARI[3]

[1]*Department of Management Studies, Graphic Era Deemed to be University, Dehradun, Uttarakhand, India*

[2]*Department of Management Studies, Adamas University, Kolkata, West Bengal, India*

[3]*University "Ukshin Hoti" Prizren, Prizren, Kosovo*

*Corresponding author

ABSTRACT

The unprecedented infusion of Artificial Intelligence (AI) and Machine Learning (ML) into Human Resource Management (HRM) has reconfigured the discourse of organizational sustainability, necessitating a conceptual re-evaluation of HR's evolving role. This study undertakes a critical analysis of HR sustainability within the technological epoch, emphasizing the interplay between human-centric values and algorithmic efficiency. It elucidates how strategic workforce planning, ethical governance of digital tools, and the cultivation of continuous learning cultures are indispensable to reconciling technological advancement with human development. The integration of AI and ML in HR functions—ranging from recruitment, onboarding, and learning to mentoring and performance evaluation—augments efficiency, accuracy, and predictive capabilities. However, it also provokes ethical dilemmas concerning data privacy,

Robo-Advisors and Artificial Intelligence in Human Resources Management: Revolutionizing HR.
Sanjay Taneja, Swati Gupta, Girish Lakhera, Mohit Kukreti, and Ercan Özen (Eds.)

DOI: 10.1201/9781779641410-7

inclusivity, and the preservation of human agency. The research argues that sustainable HR in the AI era is contingent upon fostering symbiotic human–AI collaboration, strengthening employee well-being, and institutionalizing agility in HR processes. Ultimately, the paper posits that while AI and ML catalyse transformative efficiency, HR sustainability lies in harmonizing digital capabilities with human intuition, thereby redefining organizational resilience, inclusivity, and long-term value creation in the age of intelligent systems.

7.1 INTRODUCTION

On the other hand, some claim that HR is the sole department that primarily deals with people. Since HR-related jobs are handled artistically, the existence of AI may undermine this humanistic approach. Complex challenges can be solved more permanently when the human side of business is handled skillfully (Britannica, n.d.; Biswas, 2018). The cooperation of HR and Information and Technology (IT) is not a recent development. Human resource information systems (HRIS), enterprise resource planning (ERP), the Internet of Things (IoT), HR analytics, data mining, and, most recently, artificial intelligence (AI) have all been examples of how IT has been incorporated into HR for generations on end (Bondarouk, Harms, and Lepak, 2015). We have often acknowledged how important technology breakthroughs are to the overall expansion of the company (Biswas, 2018). Managers from many fields, including HR, are drawn to the technology's simplified answers to the intricate issues facing every department within a company.

Human Resources (HR) sustainability in the age of AI and machine learning (ML) involves a conceptual analysis that considers the long-term impact on organizations, employees, and the HR function itself (Budhwar et al., 2023). Here's an exploration of key concepts in HR sustainability within this technological context:

7.1.1 HUMAN RESOURCE SUSTAINABILITY CONCEPT

1. **Strategic Workforce Planning:**
    - Concept — HR sustainability involves strategic workforce planning that adapts to the evolving landscape of AI and ML. It encompasses aligning the organization's human capital

with technological advancements to ensure a skilled, resilient workforce.

- Analysis: As AI and ML reshape job roles, HR sustainability requires proactive planning for skills development, re-skilling, and up-skilling to match the demands of emerging technologies (Charlier and Kloppenburg, n.d.).

2. **Talent Acquisition and Retention:**

- Concept— Sustainable HR practices leverage AI and ML for efficient talent acquisition, considering not only technical skills but also adaptability, creativity, and emotional intelligence.
- Analysis: Emphasizing the human-centric aspects of work becomes crucial for talent retention. Sustainable HR focuses on creating a work environment that complements the capabilities of both humans and machines (Chowdhury et al., 2023).

3. **Ethical Use of Technology:**

- Concept— HR sustainability underscores the ethical use of AI and ML, ensuring fairness, transparency, and accountability in decision-making processes.
- Analysis: Establishing ethical guidelines for the deployment of AI in HR functions prevents biases, maintains employee trust, and contributes to a positive organizational culture.

4. **Employee Well-Being:**

- Concept— Sustainable HR recognizes the impact of AI on employee well-being and strives to create a work environment that supports mental health, work-life balance, and a sense of purpose.
- Analysis: AI and ML can alleviate routine tasks, allowing HR to focus on initiatives promoting employee well-being, career development, and a sense of fulfillment (Copeland, n.d.).

5. **Continuous Learning Culture:**

- Concept— HR sustainability fosters a culture of continuous learning to adapt to technological advancements, encouraging employees to embrace up-skilling and re-skilling opportunities.

- Analysis: Leveraging AI for personalized learning plans and identifying individual development needs ensures that employees remain agile and ready for evolving job roles (Di Vaio et al., 2020).

6. **Inclusive Workforce:**
   - Concept— Sustainable HR promotes diversity and inclusion in the age of AI, recognizing the importance of diverse perspectives in developing and implementing AI technologies.
   - Analysis: AI algorithms must be designed with inclusivity in mind to prevent biases. Sustainable HR practices involve creating diverse talent pipelines and ensuring fair representation at all levels (Higginbottom, 2018).

7. **Human-AI Collaboration:**
   - Concept— HR sustainability envisions a future where humans and AI collaborate seamlessly. It emphasizes the augmentation of human capabilities through AI, rather than replacement.
   - Analysis: Cultivating a collaborative mindset among employees, where they see AI as a tool to enhance their work, is essential for HR sustainability. This involves effective communication and change management (Hooda, 2018).

8. **Agile HR Processes:**
   - Concept— Sustainable HR embraces agility in processes to respond promptly to changes in the technological landscape, organizational needs, and employee expectations.
   - Analysis: Integrating AI into HR processes allows for real-time data analysis, enabling quicker decision-making and responsiveness to evolving trends, contributing to organizational agility.

9. **Data Governance and Privacy:**
   - Concept— Sustainable HR emphasizes robust data governance practices, ensuring the responsible use of AI-generated insights while safeguarding employee privacy (Jain, 2017).

- Analysis: Striking a balance between leveraging data for informed decision-making and respecting privacy rights is crucial for building trust and maintaining ethical standards.

10. HR Leadership in Tech Adoption:

- Concept— HR sustainability involves HR leaders taking a proactive role in guiding the organization through the ethical adoption of AI and ML.
- Analysis: HR leaders must champion ethical AI practices, promote a culture of curiosity and learning, and act as advocates for employees, ensuring that technological advancements align with organizational values and goals.

7.2 LITERATURE REVIEW

7.2.1 ARTIFICIAL INTELLIGENCE

A computer-controlled robot that intelligently completes a task for a human being is a common example of AI. The main uses of AI include the phrase, which is commonly used in reference to tasks requiring a great deal of data analysis (Budhwar et al., 2023; Lawler and Elliot, 1996). Computer technology has advanced over time, offering quick fixes and reactions to corporate issues, but it has never been able to match human adaptability. Nonetheless, in broader business domains, several AI technologies have demonstrated exceptional performance, matching that of professionals and human experts (Britannica, online library).

7.2.2 MACHINE LEARNING (ML)

ML is applied in various domains, including NLP, computer vision, speech recognition, healthcare, finance, and many others (Higginbottom, 2018). The availability of large datasets and powerful computing resources has contributed to the rapid advancements in ML in recent years. Popular ML libraries and frameworks, such as TensorFlow and PyTorch, make it easier for researchers and practitioners to implement and experiment with ML models (Robinson and Sparrow, 2007).

7.2.3 MACHINE LEARNING IN HR

ML is increasingly being integrated into various aspects of HR to enhance efficiency, improve decision-making, and streamline processes. While ML offers numerous benefits to HR processes, it's crucial to address ethical considerations, data privacy, and ensure transparency in algorithmic decision-making (Meister, 2017). Additionally, human oversight is essential, especially in areas that require empathy, understanding of cultural nuances, and ethical decision-making. Integrating ML into HR should be approached with a balanced and thoughtful strategy to maximize its effectiveness and ensure responsible use.

7.2.4 AI IN HR

There is ongoing skepticism over the idea that integrating AI into HR processes could cause highly intelligent software to replace human workers in various positions. Still, there will undoubtedly be a significant amount of job redesign and reclassification at all levels (Strohmeier and Piazza, 2013). A paper by PWC titled "AI in HR: A No-brainer" suggests that certain industries possibly will experience noteworthy employment loss, while others may see the creation of original positions requiring highly digitalized intelligence. Data pertaining to HR can be obtained from the HR department of any given organization. The examination, evaluation, and interpretation of the facts are frequently carried out in order to facilitate the results of several crucial organizational decisions. AI is useful because it saves time compared to using humans to analyze exceedingly comprehensive data (Jain, 2016). According to Robinson and Sparrow (2007), AI facilitates intelligent database administration for the HR department at different levels and streamlines individual competency management, continuous monitoring, and competency assessment (Stone and Dulebohn, 2013).

7.2.5 POSITIVE VIEW OF AI IN HR

AI will increase the entire worth of HR, according to a different viewpoint on digital HR. While ML and automation will handle routine, basic jobs that require understanding of methodology, HR will concentrate more on tasks that call for problem-solving and leadership abilities. AI offers the interface necessary for the efficient performance of HR-related tasks (EY, n.d.). The potential advantages of AI in HR notwithstanding, it is imperative to apply AI in an ethical and open manner, taking privacy, bias, and managerial issues into consideration. Furthermore, it's critical to have a human touch in HR procedures to guarantee that AI use enriches and integrates personnel the management's human component.

7.2.6 POSITIVE VIEW OF MACHINE LEARNING IN HR

The integration of ML in HR has the potential to bring about numerous positive impacts, improving efficiency, objectivity, and the overall employee experience. While recognizing these positive aspects, it's important to emphasize that ML should be implemented responsibly and ethically. Human oversight, transparent communication, and continuous monitoring are essential to make certain that the expertise is used in a method that aligns with organizational standards and respects individual privacy and rights. When applied thoughtfully, ML has the potential to significantly enhance HR practices and contribute to a positive workplace environment (Unahabhokha, Platts, and Tan, 2007).

7.3 DIGITALIZATION OF HR FUNCTIONS

7.3.1 HIRING AND TALENT ACQUISITION

Hiring is the process of filling immediate job openings in an organization by identifying needs, creating job descriptions, advertising, and selecting candidates. It's a transactional and short-term focus on fulfilling specific roles (Chowdhury et al., 2023). Talent acquisition, on the other hand, is a strategic and long-term approach to identifying, attracting, and nurturing top talent for an organization. It involves strategic planning, employer branding, continuous relationship building, and sourcing from various channels. Talent acquisition extends beyond immediate needs, aiming to build a pipeline of skilled individuals aligned with the organization's long-term goals (Unahabhokha, Platts, and Tan, 2007).

7.3.2 ON-BOARDING

As soon as the new hire is brought on board, they begin a sequence of routine tasks. For the executives in charge of the HR department, directly managing a newly hired employee will take a lot of time and energy. The belief is that AI plays more of a facilitator role in the HR function, assisting HR in completing tasks faster and allocating more time to focus on learning and developing critical competencies like "design skills, psychological understanding, etc." to make critical decisions for the company (Lawler and Elliot, 1986; Lamson, 2018).

7.3.3 LEARNING AND DEVELOPMENT

According to Jain (2016), incorporating AI-enabled tools into the design of employee learning and development opportunities can improve job satisfaction. Creating career development strategies that meet the needs of the workforce reduces absenteeism and boosts productivity (Jain, 2016). AI is already being used as an erudition execute, as President and CEO of Lamsen Consult Melissa Lamson noted in a recent piece (Tambe, Cappelli, and Yakubovich, 2019). Asking Alexa or Siri for advice is nothing more than a teaching tool. These days, we learn through experiences rather than instructions via digital media. Because AI has made it possible to concentrate on experiential learning, design and critical thinking are being emphasized.

7.3.4 COACHING AND MENTORING

AI is unlikely to replace coaching and mentoring, which is another essential HR function. In contrast, programs such as "mobile coach" help individuals identify their desires and create modified learning and development campaigns (Vrontis et al., 2022). Thus, software may evaluate an employee's work requirements and provide guidance for their potential career routes and group formation. Employees now have fast access to possibilities for coaching and mentoring thanks to a multiplicity of mini-course formats. Still, there are a few delicate concerns concerning human involvement. Human involvement is still necessary to interpret situations related to employee behavior and attitudes (Hooda, 2018).

7.4 CONCLUSION

Certainly, very methodical and precise business solutions for HR have been made possible by AI and ML. AI is facilitating the complex duties like managing personnel, developing employees, evaluating them, assigning benefits, choosing employees, engaging them, monitoring their performance, and feedback, so forth. How much technology is used in HR is something that the HR managers need to determine. Researchers differ on whether AI can replace HR's vital role because AI can respond more quickly and accurately than HR, and some think AI cannot replace HR's contribution.

It is now the responsibility of HR managers to determine how much ML is allowed to be used in HR tasks, given the integration of AI, ML, and HR. HR must clearly define the boundaries between tasks handled by HR and ML-ruled activities. Additionally, HR should exert influence over AI in all roles. According to the current study, AI controls the effective implementation of HR functions after discussing the perspectives of numerous company professionals on the subject.

Even if ML has reduced labor costs and increased productivity across a range of HR domains, human intervention remains crucial for the successful execution of HR operations. The HR department alone is responsible for maintaining special associations and gathering behavioral insights into human thinking. The companionship might never be able to preserve staff members or their commitment to the firm if all of the jobs are given to the machines. In order to accomplish their current tasks more effectively, HR must enlist the help of AI and ML in addition to continuing to do so.

7.5 FUTURE RESEARCH DIRECTION

The growing use of technology in HR has been covered in a number of publications and research papers, either as HR digitization or as HR becoming more dependent on AI and ML. Though it has made an effort to address many aspects of technology utilized in HR operations, this study still offers room for further investigation. The foundation for further ML and HR research projects is laid by this study. In the future, scientists can benefit from the developing state of ML and HR to conduct empirical study and contribute to the body of knowledge in the relevant field.

Researchers can do a comparative analysis to determine how much easier labor was before and after AI and ML were used in HR operations. Additionally, some scholars may find the current study to be thought-provoking as they consider the various job descriptions in the context of AI.

KEYWORDS

- **artificial intelligence**
- **digitization**
- **HRM**
- **machine learning**
- **sustainability**

REFERENCES

Biswas, S. (2018). Employee experience, jobs and skills: How AI will impact HR. *HR Technologist*. https://www.hrtechnologist.com/articles/digitaltransformation/employee-experience-jobs-and-skills-how-ai-will-impact-hr/ (accessed on 11th August 2025).

Bondarouk, T., Harms, R., & Lepak, D. (2015). Does e-HRM lead to better HRM service? *The International Journal of Human Resource Management*, 1–31.

Britannica. (n.d.). Artificial intelligence. In *Encyclopedia Britannica*. https://www.britannica.com/technology/artificial-intelligence (accessed on 11th August 2025).

Budhwar, P., Chowdhury, S., Wood, G., Aguinis, H., Bamber, G. J., Beltran, J. R., & Varma, A. (2023). Human resource management in the age of generative artificial intelligence: Perspectives and research directions on ChatGPT. *Human Resource Management Journal, 33*(3), 606–659.

Charlier, R., & Kloppenburg, S. (n.d.). A report on artificial intelligence in HR: A no-brainer. https://www.pwc.at/de/publikationen/verschiedenes/artificial-intelligence-in-hr-a-nobrainer.pdf (accessed on 11th August 2025).

Chowdhury, S., Joel-Edgar, S., Dey, P. K., Bhattacharya, S., & Kharlamov, A. (2023). Embedding transparency in artificial intelligence machine learning models: Managerial implications on predicting and explaining employee turnover. *The International Journal of Human Resource Management, 34*(14), 2732–2764.

Copeland, B. J. (n.d.). Artificial intelligence. https://www.britannica.com/technology/artificial-intelligence (accessed on 11th August 2025).

Di Vaio, A., Palladino, R., Hassan, R., & Escobar, O. (2020). Artificial intelligence and business models in the sustainable development goals perspective: A systematic literature review. *Journal of Business Research, 121*, 283–314.

EY. (n.d.). The new age: Artificial intelligence for human resource opportunities and functions. https://www.ey.com/Publication/vwLUAssets/EY-the-new-age-artificial-intelligence-for-human-resource-opportunities-and-functions/$FILE/EY-the-new-age-artificial-intelligence-for-human-resource-opportunities-and-functions.pdf (accessed on 11th August 2025).

Higginbottom, K. (2018). *The Impact of AI on the HR Profession*. Forbes. https://www.forbes.com/sites/karenhigginbottom/2018/06/14/the-impact-of-ai-on-the-hrprofession/#7364c0e391ca (accessed on 11th August 2025).

Hooda, S. (2018). 5 ways HR will be affected by artificial intelligence. *People Matters*. https://www.britannica.com/technology/artificial-intelligence (accessed on 11th August 2025).

Jain, S. (2016). Big data analysis: Changing the way. *International Journal of Research in Advanced Engineering and Technology, 2*(6), 34–36.

Jain, S. (2017). Is artificial intelligence the next big thing in HR? Paper presented in *International Conference in Innovative Research in Science, Technology and Management*, January 22–23, 2017.

Lamson, M. (2018). The role of AI in learning and development—Make way for artificial intelligence in learning and development, it's here. https://www.inc.com/magazine/201810/maria-aspan/brene-brown-leadership-consultantresearch.html (accessed on 11th August 2025).

Lawler, J. J., & Elliot, R. (1996). Artificial intelligence in HRM: An experimental study of an expert system. *Journal of Management, 22*(1), 85–111.

Meister, J. (2017). The employee experience is the future of work: 10 HR trends for 2017. *Forbes*. https://www.forbes.com/sites/jeannemeister/2017/01/05/the-employee-experience-is-thefuture-of-work-10-hr-trends-for-2017/#3b9d7cf020a6 (accessed on 11th August 2025).

Robinson, M. A., & Sparrow, P. R. (2007). Forecasting future competency requirements: A three-phase methodology. *Personnel Review, 36*(1), 65–90.

Stone, D., & Dulebohn, J. (2013). Emerging issues in theory and research on electronic human resource management (e-HRM). *Human Resource Management Review, 23*, 1–5.

Strohmeier, S., & Piazza, F. (2013). Domain driven data mining in human resource management: A review of current research. *Expert Systems with Applications, 40*, 2410–2420.

Tambe, P., Cappelli, P., & Yakubovich, V. (2019). Artificial intelligence in human resources management: Challenges and a path forward. *California Management Review, 61*(4), 15–42.

Unahabhokha, C., Platts, K., & Tan, K. H. (2007). Predictive performance measurement system: A fuzzy expert system approach. *Benchmarking: An International Journal, 14*(1), 77–91.

Vrontis, D., Christofi, M., Pereira, V., Tarba, S., Makrides, A., & Trichina, E. (2022). Artificial intelligence, robotics, advanced technologies and human resource management: A systematic review. *The International Journal of Human Resource Management, 33*(6), 1237–1266.

CHAPTER 8

Optimizing Talent Acquisition: A Comprehensive Analysis of Recruitment and Selection Practices in the Manufacturing Industry

PRASHANT KUMAR SIDDHEY,* UTTAM KAUR, RUBEL AMIN, and K. K. MISHRA

University School of Business, Chandigarh University, Gharuan, Mohali, Punjab, India

*Corresponding author

ABSTRACT

This study interrogates the evolving architecture of recruitment and selection (R&S) practices within India's manufacturing sector, foregrounding their strategic salience for organizational competitiveness and workforce sustainability. Anchored in empirical analysis, the research evaluates how structured processes—spanning recruitment design, career growth prospects, promotion mechanisms, and clarity in training frameworks—collectively shape employee perceptions and organizational outcomes. Data were collected from 100 respondents, comprising general and line managers across Madhya Pradesh, and analyzed using descriptive statistics, factor analysis, and reliability measures via SPSS 20.0. The findings delineate four critical determinants of R&S effectiveness: recruitment process, future career trajectories, advancement policies, and transparency in rules and training opportunities. Intriguingly, statistical results reveal negligible variation across organizational size, suggesting that R&S practices in manufacturing are relatively standardized, irrespective of scale.

Robo-Advisors and Artificial Intelligence in Human Resources Management: Revolutionizing HR. Sanjay Taneja, Swati Gupta, Girish Lakhera, Mohit Kukreti, and Ercan Özen (Eds.)

DOI: 10.1201/9781779641410-8

While traditional methods such as internal promotions and employee referrals remain dominant, the study underscores the imperative of integrating contemporary approaches, including artificial intelligence and augmented reality, to enhance precision, inclusivity, and resilience. The research posits that future-ready R&S systems must harmonize established practices with technological innovations, thereby strengthening human capital strategies and ensuring sustainable growth in the manufacturing sector.

8.1 INTRODUCTION

Human resource management (HRM) is a vital component in determining an organization's course in the dynamic world of business (Mahapatro, 2022). The cornerstones of HR are efficient recruiting and selection (R&S) procedures, which guarantee that the appropriate people are in the right jobs and ultimately propel productivity, creativity, and organizational success (Mphaphuli, 2022). Human capital is very important and cannot be understated. The personnel of a firm are the backbone of its operations, the inspiration behind its successes, and the cornerstone of its future development (Batool et al., 2023). HR experts' main duty is to harness this human energy and direct it toward organizational goals (Urme, 2023).

8.1.1 THE EVOLUTION OF RECRUITMENT PRACTICES

The workforce is always changing, and recruitment procedures have had to change with it (Gardner, 2021). The search for outstanding talent has fueled the constant improvement of recruitment tactics since the Roman army's early days of employee recommendations through today's complex talent management systems. The usefulness of employee referrals in locating and attracting skilled individuals is demonstrated by the Roman army's use of them (Wren and Bedeian, 2023). This strategy has shown to be an effective means of locating high-potential people since it makes use of the networks and referrals of current workers (Singh and Shaurya, 2021).

8.1.2 THE CHANGING LANDSCAPE OF RECRUITMENT

Organizations and their employees need to put in more than just hard effort in the ever-evolving and fast-paced business world (Gupta et al., 2022). Companies must embrace innovation and adjust to the changing external environment if they want to succeed in the cutting edge of competitiveness (Raveendra, Satish, and Singh, 2020). The success of a company depends on the hiring, training, and retaining of competent labor. The secret to maximizing a company's workforce's potential is an efficient R&S process (Cutshall et al., 2021).

8.1.3 THE IMPORTANCE OF A STRUCTURED PROCESS

Organizations may be sure to find and hire the best applicants for their open positions by using a systematic R&S procedure (Vardarlier and Zafer, 2020). There are usually multiple steps in this procedure, which include:

1. **Job Examination and Specification:** To identify the ideal applicant profile, a systematic examination of the job's responsibilities, duties, and necessary skills and credentials is carried out (Vernau et al., 2021).
2. **Development of Recruiting Strategies:** Taking into account the target talent pool and the job needs, the best channels for recruiting are determined. Online "job boards," "social media platforms," "employee referrals," or "professional networks" are a few examples of these channels (Kamran et al., 2015).
3. **Candidate Sourcing and Screening:** To create a shortlist of applicants who fit the requirements, potential candidates are found and vetted using information from their web profiles, cover letters, and resumes (White-Lewis, 2020).
4. **Assessment and Evaluation:** To further determine whether a candidate is qualified for the position, a battery of tests and interviews, including psychometric evaluations, are conducted on the chosen applicants (Thompson et al., 2023).
5. **Reference Checks:** To confirm a candidate's credentials, experience, and employment history, reference and background checks are made (Ramadhani and Septemuryantoro, 2023).
6. **Selection and Offer:** After careful consideration, the best qualified applicant is given the job offer along with a detailed employment contract that details the terms and conditions of employment (Abbas, Shah, and Othman, 2021).

8.1.4 ADAPTING TO CHANGING NEEDS

R&S procedures need to change to keep up with the workforce's altering needs and the industries' changing requirements. Organizations must take into account elements like technological advancements: The need to find applicants with the requisite technical abilities and digital literacy is driven by the growing influence of technology in the workplace (Smaldone et al., 2022). Multidisciplinary Skills: Organizations must look for personnel that can adapt to a variety of jobs and responsibilities due to the growing demand for multidisciplinary skills (Dwivedi et al., 2023). Virtual Work Environments: Because remote and hybrid work arrangements are so common, it is imperative to evaluate candidates' comfort levels in virtual work environments (Hopkins and Bardoel, 2023). Evolving Skill Requirements: The rapid pace of technological and organizational change requires organizations to stay abreast of emerging skills and adapt their recruitment strategies accordingly (Gouda, 2022).

8.1.5 RESEARCH QUESTIONS

1. What are the primary factors that influence R&S practices in India's manufacturing sector?
2. How does company size influence the prioritization and implementation of factors affecting R&S practices?
3. What strategies can be implemented to enhance the effectiveness of R&S practices in India's manufacturing sector?

Thus, in India's dynamic manufacturing sector, this study on R&S practices is essential for navigating challenges and capitalizing on opportunities in acquiring skilled individuals. As a cornerstone of the nation's economy, manufacturing drives growth, employment, and technological advancements. Focused on key stakeholders like general and line managers in Madhya Pradesh, the research provides a regional perspective, uncovering diverse recruitment practices. The study's significance lies in identifying prevalent methods, comprehensively understanding the current landscape, aligning strategies with employee expectations, and proposing measures for enhanced practices. It captures diverse perspectives through questionnaires and statistical analysis, revealing a crossroads where traditional and innovative methods coexist. Advocating for embracing technologies like augmented reality (AR) and AI, the study serves as a guide for a more competitive and sustainable future in India's manufacturing sector.

8.2 REVIEW OF LITERATURE

The adoption of staffing practices varies significantly across various sectors. As per research, industries such as information technology and engineering are more inclined towards implementing technology-enriched recruitment methods, whereas the manufacturing industry still prefers less technological approaches (Cutshall et al., 2021; Fein, 1998). The primary focus of staffing practices within an organization is to match the capabilities of potential candidates with the job's demands (Batool et al., 2023; Montgomery, 1996).

Manufacturing firms commonly utilize internal sources of recruitment, such as promotions, transfers, or job rotations. The typical selection steps for choosing managers in Malaysia's industry include a scrutiny interview and resumes, technical and HR interview, and a final selection (Thompson et al., 2023; Jusoh and Kokai, 2002).

Internet recruiting tools have been steadily increasing (Anderson and Witvliet, 2008). R&S are crucial to an organization's success, and a knowledgeable and skilled workforce is a valuable asset (Khang, Jadhav, and Birajdar, 2023; Khandekar and Sharma, 2005).

The quality of applicants recruited in the first stage of the staffing process is critical to an organization's success (Ployhart, 2006). Whereas, online methods using audio and video are more suitable for tech-oriented workers and less appropriate for labor jobs (Parry and Tyson, 2008; AhmedSchel and Schroeder, 2002).

Employee behavioral traits can significantly impact the success of total quality management (Akanmu et al., 2023). Quality management practices positively influence overall competitiveness, and management should closely consider potential employees' behavioral traits (Taylor and Bergman, 1978).

The effectiveness of a five-stage recruitment program is intricately linked to the comparability of job offers, work experience, and opportunities (Uchiyama et al., 2022). In a field setting, researchers employed a correlational design to underscore the program's effectiveness, underscoring the pivotal role these factors play in the recruitment process (Arvey and Gordon, 1975).

The temporal gap between the initial psychological testing can serve as a noteworthy impediment to the hiring process (Scerpella et al., 2022). The examination of the recruitment timeline brought attention to the issue of time lag in the selection process (Alderfer and McCord, 1970).

A study delved into the impact of interviewer behaviors and attitudes on job candidate interview evaluations (Watson et al., 2018). Analyzing a sample of 112 master's degree students, the research unveiled that candidates with advanced needs perceive interviewers as more interested and concerned. The study underscores the necessity of refining recruitment strategies to ensure effective completion of recruitment schedules (Rothstein and Goffin, 2006).

Exploring the reasons behind the surge in popularity of personality assessments in personnel selection, researchers found that internet-based assessments address the issue of candidates faking their personality tests during selection (Chen and Chen, 2008). Proposing a data mining framework utilizing decision trees and association rules for personnel selection, their approach aids organizations in identifying key factors affecting the selection process. The study's outcomes, coupled with expert discussions, contributed to the development of specific recruitment and HRM strategies (Holm, 2012).

An investigation by an author focused on the influence of e-recruitment on the hiring process, suggesting its potential to transform traditional hiring into a more collaborative and time-independent process (Wu and Kao, 2022).

For manufacturing companies, finding candidates with the right technical skills poses the most significant recruiting challenge, as indicated by a survey among HR professionals, with 63% identifying this issue among their top three challenges (Divyalakshmi et al., 2017).

Recruitment, according to Kamran et al. (2015), involves acquiring the right people to sustain a company's operations. It encompasses creating a pool of potential candidates to fill open positions, and staff selection is a crucial part of the recruitment process, involving the selection of competent and qualified applicants. Kamran highlighted issues associated with different R&S techniques used by various companies, examining the anticipated process outcomes.

Emphasizing the importance of establishing effective R&S methods, the narrative underscores their role in saving time and costs associated with searching, interviewing, hiring, and training employees. Furthermore, it highlights the pivotal role of proper recruitment in enabling line managers to effectively work towards achieving the organization's primary objectives.

8.3 OBJECTIVES OF THE STUDY

The main objective of this study is to examine the R&S techniques being practiced in the manufacturing firms of India. The following is the set of sub-objectives:

1. To study factors affecting R&S in India's manufacturing sector.
2. To compare the significance of factors affecting R&S in India's manufacturing sector based on company size.
3. To suggest measures to improve R&S in India's manufacturing sector.

8.3.1 HYPOTHESES OF THE STUDY

The following alternative hypotheses have been formulated to accomplish the third objective of the study:

- **Hypothesis H11:** There is a significant difference concerning the Recruitment Process on R&S in the Manufacturing sector of India based on company size.
- **Hypothesis H12:** There is a significant difference regarding future career growth in R&S in India's manufacturing sector based on company size.
- **Hypothesis H13:** Company size There is a significant difference concerning promotion and career advancement policies on R&Sin India's manufacturing sector based on company size.
- **Hypothesis H14:** There is a significant difference between Rule Clarity and Training Opportunities on R&S in India's manufacturing sector based on company size.

8.4 RESEARCH METHODOLOGY

In this research, a descriptive methodology was employed to collect and analyze primary data through the use of questionnaires. The study focused on manufacturing industries in Madhya Pradesh, India, with the target population consisting of general managers and line managers. A total of 100 respondents participated and completed the questionnaires. The data analysis was conducted using SPSS software version 20.0. Prior to applying factor analysis, the collected data underwent Bartlett's Test of Sphericity and Kaiser-Meyer-Olkin (KMO) measures. Bartlett's Test of Sphericity indicated statistically significant correlations among the variables (chi-square = 190.741, df = 91). The Kaiser-Meyer-Olkin value was 0.715, indicating that the data were suitable for factor analysis—a descriptive study involving primary data collection and analysis via questionnaires. A representative sample of 100 participants from Indore and nearby areas was targeted. The data was collected from GM/AGM/HRM personnel and employees using Google Forms. Initially, 26 statements were included in the questionnaire. Following expert review, only 21 words with a 75% or higher score were retained for further analysis.

Factor Analysis: Factor analysis (KMO = 0.715) was applied to the responses. Five statements with factor loadings below 0.4 were eliminated, resulting in four factors:

- F1: Recruitment Process
- F2: Future Career Growth
- F3: Promotion and Career Advancement Policies
- F4: Rule Clarity and Training Opportunities

8.4.1 RELIABILITY

The alpha coefficient (Cronbach's alpha) was calculated at 0.702. Statistical techniques employed, including Cronbach's alpha, KMO, factor analysis, mean, standard deviation, percentage analysis, and ANOVA, were computed using SPSS 20.0.

8.4.2 QUESTIONNAIRE DEVELOPMENT

The data analysis utilized the SPSS software package, version 20.0. Before conducting factor analysis, the data underwent testing using Bartlett's Test of Sphericity and KMO measures. Bartlett's Test of Sphericity demonstrated statistically significant correlations among the variables, with an approximate chi-square of 190.741, df of 91, and a significance level of .000. The Kaiser-Meyer-Olkin measure of sampling adequacy, with a recorded value of 0.715, exceeded the threshold of 0.60, indicating the sample's appropriateness for factor analysis. These criteria collectively affirmed the suitability of the data for factor analysis. For further analysis, only factor loadings exceeding 0.40 were considered.

These factors were named Recruitment Process, Future Career Growth, Promotion and Career Advancement Policies, Rule Clarity, and Training Opportunities. The reliability coefficient of alpha (Cronbach's alpha) was 0.702.

Initially, 26 statements were asked in the questionnaire. After checking its validity through expert review, only 21 words that scored more than 75% or were in the fourth quartile were taken for further study. Then, factor analysis (KMO = 0.715) was applied to the responses. Five statements had a low factor load of less than 0.4. The remaining statements were grouped into four factors:

- F1: Recruitment Process.
- F2: Future Career Growth.
- F3: Promotion and Career Advancement Policies.
- F4: Rule Clarity and Training Opportunities.

The reliability coefficient of alpha (Cronbach Alpha) was 0.801.

8.5 RESULTS AND ANALYSIS

Table 8.1 shows that the respondents were 7% from the small-scale industry, 45% from the medium size, and 48% from the large size. 46% of respondents were male, and 54% were female. 63% of respondents were in the age group of 20–30, 27% were in the age group of 30–40, and 10% were in the age group of 40–50. 10% of respondents were just metric pass, 20% were undergraduate, 30% were postgraduate, and 40% were above postgraduate. 17% of respondents had experience of up to 5 years, 33% had experience between 5–10 years, and 15% had experience between 10–15 years.

TABLE 8.1 Descriptive Information

		N	Mean	Std. Deviation
F1	SMALL	7	3.67	0.38
	MED	45	3.21	0.65
	LARGE	48	3.21	0.75
	Total	100	3.24	0.69
F2	SMALL	7	2.74	0.90
	MED	45	2.72	0.93
	LARGE	48	2.76	1.15
	Total	100	2.78	1.04
F3	SMALL	7	3.51	0.76
	MED	45	3.26	0.79
	LARGE	48	3.34	0.83
	Total	100	3.32	0.80
F4	SMALL	7	4.36	0.48
	MED	45	4.16	0.79
	LARGE	48	4.28	0.77
	Total	100	4.30	0.76

Note: F1: Recruitment Process. F2: Future Career Growth. F3: Promotion and Career Advancement Policies. F4: Rule Clarity and Training Opportunities

TABLE 8.2 Hypotheses Testing Results

SL. No	Hypothesis	p Value	Significance	Action
1.	H_1	0.254	5%	R
2.	H_2	0.845	5%	R
3.	H_3	0.627	5%	R
4.	H_4	0.771	5%	R

Table 8.2 shows that the hypothesized results indicate that, in India's producing industry, different parts of the R&S process are not greatly impacted by the size of the organization. To be more precise, there is no discernible difference in the strategies used by large and small businesses when it comes to hiring, future prospects for professional growth,

promotion and advancement plans, rule clarity, or training possibilities. This suggests that career development strategies and recruitment and selection procedures are standard across various sizes of businesses in this industry. Regardless of the size of the manufacturing organization, employees can anticipate the same procedures and chances for professional advancement. This consistency shows that recruitment and selection procedures and industry norms are well-established and universally followed, creating an even playing field for all workers in the production sector.

8.6 CONCLUSION AND DISCUSSION

The basic idea of the research was to study existing R&S practices in the manufacturing industry. The four factors that emerged from the study are the Recruitment Process, Future Career Growth, Promotion and Career Advancement Policies, Rule Clarity, and Training Opportunities.

An extensive literature review was done to accomplish the first objective of the study, and 26 parameters were identified, which were further reduced to 21 after checking their validity. Then, managers and employees were approached by various manufacturing industries to find out existing R&S practices.

A structured questionnaire was used to accomplish the second objective of the study, which was answered by the employees of the organization about their perception of the importance of various R&S techniques in the manufacturing industry.

Various tests were applied to accomplish the third and fourth objectives, such as KMO, Factor Analysis, Mean, Standard deviation, Percentage Analysis, and ANOVA, which were calculated using SPSS 20.0 to find out the perception of employees towards the importance of R&S practices in the organization. Also, some suggestive measures were given based on the study's results.

The four factors that emerged from the study are the Recruitment Process, Future Career Growth, Promotion and Career Advancement Policies, Rule Clarity, and Training Opportunities. It has been concluded from this recruitment process project that the most commonly used recruitment methods in manufacturing organizations are employees from within, promotions, colleges and universities, so forth. Surprisingly, manufacturing industries have shown less tendency to use multiple recruitment methods, that is, temporary staffing, headhunting, and advertising in newspapers.

From the study, it can be concluded that manufacturing industries often practice employee referrals, promotion schemes, campus drives, so forth. Interestingly, manufacturing companies have shown little tendency to use specific recruitment methods, such as temporary staffing, headhunting, and newspaper advertising. Instead, they rely on traditional recruitment methods, such as promoting from within and recruiting from colleges and universities. These will likely hold significance in the future.

However, manufacturing organizations must practice new ways of R&S. During the COVID-19 pandemic, it was observed that many organizations started working from home, but it seems difficult to practice by manufacturing organizations where physical intervention of the employees is required. Although, by using AR and artificial intelligence (AI) with some kind of critical mechanics, this intervention can be minimized, this will open up a new vista for skill enhancement, and also the methods to check the skills of the prospective employees during R&S, and further study can be conducted to measure the efficiency of new ways of R&S.

KEYWORDS

- **artificial intelligence**
- **human capital**
- **manufacturing**
- **recruitment**
- **selection**
- **sustainability**

REFERENCES

Abbas, S. I., Shah, M. H., & Othman, Y. H. (2021). A critical review of recruitment and selection methods: Understanding the current practices. *Annals of Contemporary Developments in Management & HR (ACDMHR), 3*(3), 46–52.

AhmedSchel, & Schroeder, G. R. (2002). The importance of the recruitment and selection process for the sustainability of total quality management. *International Journal of Quality and Reliability Management, 19*(5), 540–550.

Akanmu, M. D., Hassan, M. G., Ibrahim Alshuaibi, M. S., Ibrahim Alshuaibi, A. S., Mohamad, B., & Othman, A. (2023). The mediating role of organizational excellence between quality management practices and sustainable performance. *Total Quality Management & Business Excellence, 34*(9, 10), 1217–1242.

Alderfer, C. P., & McCord, C. G. (1970). Personal & situational factors in recruitment interviews. *Applied Psychology, 54*(4), 377–385.

Anderson, N., & Witvliet, C. (2008). Fairness reactions to personnel selection methods: An international comparison between The Netherlands, The United States, France, Spain, Portugal, and Singapore. *International Journal of Selection and Assessment, 16*(1), 1–13.

Arvey, D. R., & Gordon, E. M. (1975). The different dropout rates of minority and majority job candidates due to the time lag between selection process. *Personnel Psychology, 28*(5), 175–180.

Batool, M., Nadim, M., Saeed, A. F., Wadood, F., Quresh, M., & Pucelj, M. (2023). Human resource strategy: Enhancing organizational performance through effective HRM. *PalArch's Journal of Archaeology of Egypt/Egyptology, 20*(2), 1556–1573.

Chen, C. F., & Chen, L. F. (2008). Data mining to improve personnel selection & enhance human capital: A study on high-technology industry. *Expert Systems with Applications, 34*(1), 280–290.

Cutshall, H., Hattaway, R., Singh, N. P., Rais-Bahrami, S., & McCleskey, B. (2021). The #Path2Path virtual landscape during the COVID-19 pandemic: Preparing for the 2020 pathology residency recruitment season. *Academic Pathology, 21*(7), 8.

Divyalakshmi, N., et al. (2017). A study on recruitment and selection process in the manufacturing industry, *3*(4), 3341–3350.

Dwivedi, Y. K., Kshetri, N., Hughes, L., Slade, E. L., Jeyaraj, A., Kar, A. K., & Wright, R. (2023). "So, what if ChatGPT wrote it?" Multidisciplinary perspectives on opportunities, challenges, and implications of generative conversational AI for research, practice, and policy. *International Journal of Information Management, 71*, 102642.

Fein, R. (1998). Traditional or electronic tools: How do people get hired? *Journal of Career Planning & Employment, 58*(4), 40–43.

Gardner, P. (2021). *Recruiting Trends, 2021–2022*. Collegiate Employment Research Institute. USA.

Gouda, H. (2022). Exploring the effects of learning abilities, technology, and market changes on the need for future skills. *Higher Education, Skills and Work-Based Learning, 12*(5), 900–913.

Gupta, S., Arthur, L. G., Chandler, N., Danielson, P., Downard, C., Ehrlich, P., & Mak, G. (2022). Is the changing landscape of fellowship recruitment during COVID-19 here to stay? *Journal of Pediatric Surgery, 57*(10), 445–450.

Holm, A. (2012). E-recruitment: Towards ubiquitous recruitment process & candidate relationship management. *German Journal of Human Resource Management, 26*(3), 241–259.

Hopkins, J., & Bardoel, A. (2023). The future is hybrid: How organizations are designing and supporting sustainable hybrid work models in post-pandemic Australia. *Sustainability, 15*(4), 3086.

Jusoh, M., & Kokai, C. (2002). Hiring managers in the manufacturing industry of Malaysia. *Journal of Transnational Management Development, 6*(3, 4), 21–38.

Kamran, A., et al. (2015). Analysis of the recruitment and selection process (pp. 1360–1364).

Khandekar, A., & Sharma, A. (2005). Organizational learning in Indian organizations: A strategic HRM perspective. *Journal of Small Business and Enterprise Development, 12*(2), 211–226.

Khang, A., Jadhav, B., & Birajdar, S. (2023). Industry revolution 4.0: Workforce competency models and designs. In *Designing workforce management systems for Industry 4.0* (Vol. 5, Issue 7, pp. 11–34). CRC Press.

Mahapatro, B. B. (2022). *Human Resource Management*. PG Department of Business Management. India.

Montgomery, C. (1996). Organization fit is key to job success. *HR Magazine, 41*(1), 94–97.

Mphaphuli, T. K. (2022). The effect of recruitment and selection on organizational performances: A case of Vhembe District Municipality (Doctoral dissertation).

Parry, E., & Tyson, S. (2008). An analysis of the use and success of online recruitment methods in the UK. *Human Resource Management Journal, 18*(3), 257–274.

Ployhart, R. E. (2006). Staffing in the 21st century: New challenges and strategic opportunities. *Journal of Management, 32*(6), 868–897.

Ramadhani, T. A., & Septemuryantoro, S. A. (2023). Strategy of the human resources department in recruiting quality employees at Quest Prime Pemuda Hotel in Semarang. In *The International Conference on Education, Social Sciences and Technology (ICESST), 2*(2), 23–32.

Raveendra, P. V., Satish, Y. M., & Singh, P. (2020). Changing landscape of recruitment industry: A study on the impact of artificial intelligence on eliminating hiring bias from recruitment and selection process. *Journal of Computational and Theoretical Nanoscience, 17*(9–10), 4404–4407.

Rothstein, M. G., & Goffin, R. D. (2006). The use of personality measures in personnel selection. *Human Resource Management Review, 16*(2), 155–180.

Scerpella, T. A., Spiker, A. M., Lee, C. A., Mulcahey, M. K., & Carnes, M. L. (2022). Next steps: Advocating for women in orthopedic surgery. *JAAOS-Journal of the American Academy of Orthopedic Surgeons, 30*(8), 377–386.

Singh, A., & Shaurya, A. (2021). Impact of artificial intelligence on HR practices in the UAE. *Humanities and Social Sciences Communications, 8*(1), 1–9.

Smaldone, F., Ippolito, A., Lagger, J., & Pellicano, M. (2022). Employability skills: Profiling data scientists in the digital labor market. *European Management Journal, 40*(5), 671–684.

Taylor, M. S., & Bergman, J. T. (1978). Organizational recruitment activities & applicant's reactions at different stages of the recruitment process. *Personnel Psychology, 40*(2), 261–285.

Thompson, I., Koenig, N., Mracek, D. L., & Tonidandel, S. (2023). Deep learning in employee selection: Evaluation of algorithms to automate the scoring of open-ended assessments. *Journal of Business and Psychology, 38*(3), 509–527.

Uchiyama, Y., Furuoka, F., & Akhir, M. N. M., (2022). Gig workers, social protection and labor market inequality: Lessons from Malaysia. *Jurnal Ekonomi Malaysia, 56*(3), 165–184.

Urme, U. N. (2023). The impact of talent management strategies on employee retention. *International Journal of Science and Business, 28*(1), 127–146.

Vardarlier, P., & Zafer, C. (2020). Use of artificial intelligence as a business strategy in the recruitment process and social perspective. In *Digital Business Strategies in Blockchain Ecosystems: Transformational Design and Future of Global Business* (Vol. 25, Issue 4, pp. 355–373).

Vernau, J., Bishop, C., Chavda, S., Weldon, A., Maloney, S., Pacey, R., & Turner, A. N. (2021). An analysis of job description and person specifications to define the requirements for obtaining employment within strength and conditioning. *Professional Strength and Conditioning, 15*(8), 7.

Watson, N. L., Mull, K. E., Heffner, J. L., McClure, J. B., & Bricker, J. B. (2018). Participant recruitment and retention in remote eHealth intervention trials: Methods and lessons learned from a large randomized controlled trial of two web-based smoking interventions. *Journal of Medical Internet Research, 20*(8), e10351.

White-Lewis, D. K. (2020). The façade of fit in faculty search processes. *The Journal of Higher Education, 91*(6), 833–857.

Wren, D. A., & Bedeian, A. G. (2023). *The Evolution of Management Thought*. John Wiley & Sons.

Wu, A. C., & Kao, D. D. (2022). Mapping the sustainable human-resource challenges in Southeast Asia's fintech sector. *Journal of Risk and Financial Management, 15*(7), 307.

CHAPTER 9

Digital Transformation in Wealth Management: Exploring How Robo-Advisors Moderate Behavioral Biases and Shape Investment Decisions

VIVEK VERMA,[1,*] PRAVEEN SINGH,[1] and PAWAN KUMAR[2]

[1]*Department of Management Studies, Graphic Era Deemed to be University, Dehradun, Uttarakhand, India*

[2]*Chandigarh University, Gharuan, Mohali, Punjab, India*

*Corresponding author

ABSTRACT

The accelerating digital transformation within wealth management has precipitated a paradigmatic shift from conventional advisory models to algorithm-driven, democratized investment platforms. Central to this disruption are robo-advisors, which amalgamate data analytics, artificial intelligence, and behavioral finance to mitigate the cognitive and emotional biases that historically compromised investor rationality. This study explores the multifaceted role of robo-advisors in attenuating biases such as overconfidence, herd mentality, confirmation bias, and loss aversion, thereby fostering disciplined, evidence-based investment behavior. By offering objective risk assessments, goal-oriented strategies, systematic rebalancing, and diversified asset allocations, robo-advisors engender greater transparency, inclusivity, and personalization in financial decision-making. The research further underscores how robo-advisors leverage behavioral nudges and digital coaching

Robo-Advisors and Artificial Intelligence in Human Resources Management: Revolutionizing HR.
Sanjay Taneja, Swati Gupta, Girish Lakhera, Mohit Kukreti, and Ercan Özen (Eds.)

DOI: 10.1201/9781779641410-9

mechanisms to counteract irrational tendencies, ultimately empowering investors to align portfolios with long-term objectives while resisting myopic market impulses. Additionally, the paper delineates the scalability advantages and efficiency gains robo-advisors confer upon wealth management institutions, enabling a seamless balance between automation and human advisory empathy. The findings elucidate that robo-advisors not only reconfigure the dynamics of investment decision-making but also redefine financial accessibility, positioning themselves as indispensable catalysts for sustainable, technology-enabled wealth creation in the digital era.

9.1 INTRODUCTION

The investment landscape is ever-changing, and in the current rapidly evolving space of wealth management and investing, we see a defining force: digital transformation, which has completely redesigned how investors interact with the complexities of financial markets. The transformation is not only a byproduct of technological progress but also an indication that investor preferences have changed, and the demand for personalized, efficient, and transparent financial services is on the ultimate rise. One of the key enablers of this digital revolution is the plethora of Robo advisors—online investment platforms that use algorithms and analytics to provide financial advice and automatic execution in managing investment portfolios. Some of the significant wealth management services include asset management, tax planning, offshore services, retirement planning, financial and legal advice, so forth, (Ślązak, 2018). In the meantime, while we await the collation of Viewpoint Review by the Study Group on a worldwide net worth, prospects for the segment have enigma eggshells gliding by Dziawgo (2021).

Traditionally, the labor-intensive decision-making systems and human advisors in wealth management have been criticized for various inefficiencies, biases, and exclusivity. After all, investors are only human and, as such, likely to make mistakes, whether it is being influenced by their emotions in times of market stress or experiencing cognitive failure in assessing risks and how to manage their portfolios. Until now, the rise of Robo advisors has appeared almost out of nowhere—like a disruptive force behind the scenes, an evolutionary leap that mirrors much-needed societal advances like blending technology with behavioral finance to bridge gaps in the flawed traditional wealth management model.

In this chapter, we consider the broad implications of the digital transformation of wealth management and how Robo-advisors help to temper our behavioral biases in framing investment decisions. We also delve into how Robo-advisors leverage technology to provide unbiased, input-agnostic investment advice (independently from human nature), which empowers investors to make more rational decisions aligned with their financial goals and objectives.

We look under the hood of Robo advisors to uncover the technology, delve into behavioral finance, and illustrate how they are transforming the wealth management industry (opinion: there is no going back) that is disrupting legacy asset managers, benefiting investors with a more efficient, affordable, and transparent solution. This report leverages empirical evidence, case studies, and real-world examples to explore how Robo advisors are altering investor behavior, engagement strategies, portfolio management efficacy patterns, and the futurity of wealth management in the digital age.

9.2 BEHAVIORAL BIASES IN INVESTMENT DECISION MAKING

Ullah and Elahi (2014) said that the world of investment decision-making is not only governed by rational analysis and objective assessments of risk and return. However, more often than not, human psychology instead takes the wheel, which brings a multitude of cognitive and emotional biases that result in misled judgment and poor decisions. These behavioral biases, which are deeply rooted in human psychology, manifest themselves in multiple ways and have the potential to significantly influence investment decisions. Digital technology (e.g., AI and consumer data) is embedded in the marketing and operational strategies of modern banking and finance segments (Quach et al., 2022).

Overconfidence: One of the most stubborn peccadilloes in this sailing pool called stock investing is overconfidence, where investors tend to believe that they know better or possess flexibility that allows them to anticipate market movements accurately (Kumar and Goyal, 2015). Years of research into the nature of overconfidence has demonstrated that being excessively confident in our own abilities can cause us to underestimate risks and trade more frequently, make poorer decisions regarding diversification, and eat away at our investment returns over time.

Another common bias is loss aversion (investors feel much more pain from losing money than the joy of winning a similar amount). That discomfort with losses causes most investors to hang on to losing investments way too long (which leads to an overall lagging performance) in order to build back their abraded capital, instead of taking early hits and moving fresh chips to the green felt.

Confirmation bias is another common mistake that investors make, wherein they tend to overlook information that contradicts their baked-in assumptions or biases. However, this bias can lead investors to make poor decisions and prevent them from adapting their strategies or perspectives according to new information.

Similarly, herd mentality will make investors more inclined to follow not their judgment but peer pressure. When droves of investors all flock to popular assets or rush for the exits at perceived risks, this herd mentality can serve to amplify market volatility and create opportunities for callous contrarian investors. This can also cause market bubbles and crashes.

It has many other biases stacked on top of it (e.g., anchoring, recency bias, framing effects), making clear judgment even more difficult and questionable due to the way in which information is framed, causing perceived risks and rewards in our mind that are actually not real. Not only do such biases crop up among individual investors, but they also surface in the context of institutional decision-making, governing the behavior of fund managers, analysts, and other market participants.

These behavioral biases form the core of investment decisions and thus portfolio outcomes; therefore, it is indispensable for both investors and wealth managers to acknowledge and understand them. Recognizing the impact of these biases and putting practices in place to minimize them allows investors to make more disciplined, informed, and rational investment decisions that are aligned with their long-term financial objectives, most importantly. This recognition can then be linked to understanding how Robo advisors may act as behavioral modification units and serve as a means of influencing investors in the direction of more scientific-based, rather than emotionally driven, investment strategies (as will be discussed further in following segments).

9.3 ROBO ADVISORS IN WEALTH MANAGEMENT CONFIGURATION

The rise of Robo-advisors has led to a massive shift in the traditional wealth management industry over the last few years. These digital platforms are distinctly different from traditional models of wealth management and have taken several steps toward financial planning and investment management through new-age methods supported by technology. In the ever-evolving ecosystem of the financial services landscape, balancing tradition and innovation, Robo-advisors have gained prominence, taking shape as a potential market disruptor that is redefining the dimensions of wealth management and transforming the way investors are maneuvering through the investment decision-making process (Singh et al., 2019). With a growing body of literature from behavioral economics and related fields, we believe that Robo-advisory services can enhance investment autonomy and decision-making (Silva, 2019). In the end, when put in contrast to generic (conventional) advising solutions offered by fund managers, Robo-advisors have the potential of helping investors reduce their behavioral biases (Uhl and Rohner, 2018).

RoboQvest is based on a Robo-advisor that is characterized by digital platforms with interactive and smart user assistance (Jung et al., 2019). This service leads clients via a process of automated financial consulting performed using information technology. Robo-advisors were born out of the early 21st century when computing power and data analytics completely transformed our ability to process big data with algorithms that can take action on it in a fraction of a second (Huang et al., 2022). Robo-advisors work rapidly and independent of geographic boundaries and time limits as they transgress into the digital sphere, where access is omnipresent since modern society increasingly operates as a 24-hour society (Huang et al., 2022). Roaring? Given the demographic tidal wave of Boomers retiring at the rate of 10,000 a day along with the Millennial generation approaching peak earning ages, it is no wonder that Robo-advisors have risen to deliver a sleek and cost-effective investment solution. Traditionally, full-service wealth management was mostly reserved for high-net-worth investors who could pay the high costs of human financial advisors. Foundational to this exclusivity was a lack of access to professional financial advice and portfolio management services left off the proverbial table for most everyday investing individuals. Today, automation is an integral part of the

21st-century lifestyle. We see that automation has become a part of almost every corporate domain and is now just stepping into the world of financial decision making (Abraham et al., 2019). Factors such as the fee associated with the service (pricing) and trustworthiness of the Robo-advisor influence the choice of Robo-advisors (Salo, 2017).

This is where the innovative Robo-advisors come in—offering automated investment processes and significantly lower costs, they have democratized wealth management through technology. Robo-advisors operate on a digital platform and analyze an investor's financial goals, risk appetite, and time horizon with algorithms and data analytics, while human advisors do not act with them. This automated process enables Robo-advisors to recommend customized investment portfolios and assemble diversified types of investments tailored to each individual investor, at a fraction of the price full-service advisors would charge.

Robo advisors also provide investors with an efficient, user-friendly investing experience, employing simple interfaces and interactive tools to guide you through the investment process. Investors can check their investment accounts, track how their portfolios are doing, and follow what is happening in the markets in real time using online platforms and mobile applications. Since the platform is user-friendly, it caters to a wide range of investors, especially millennials and digital natives who are used to ease of use in their financial services.

Transparency and objectivity of investment decision-making is one more reason why investors are taking advice from a Robo-advisor. Because Robo-advisors function on algorithms instead of human intuition, they do not discriminate based on experiences or personal biases. The objectivity it provides is one of the reasons why investors trust Robo-advisors and are open to being recommended investments by a robot.

In addition, Robo advisors provide wealth management firms with a scalable answer to increasing an account base and streamlining operations. Robo advisors put much of this work on autopilot by automating the mundane tasks of portfolio recalibration, tax management, and client onboarding so that wealth managers can focus on what they do best: financial planning, client relationships, and business strategy. That scalability ultimately makes Robo advisors more attractive to wealth managers who want or need to react quicker than a custom-built solution will allow.

The result: Robo-advisors (smart software platforms that simplify investment and financial planning), powered by technology, are democratizing advice, creating transparency and efficiency in the wealth management industry, ultimately leading up to the era of data-driven democratization of wealth management.

9.4 HOW BEHAVIORAL BIASES ARE KEPT IN CHECK BY ROBO ADVISORS

Robo advisors, which serve as slight counters to investors' behavioral biases, employ high-end technology and data-driven strategies to further ensure that investors make dispassionate investments. These automated platforms are designed to mitigate the impact of the cognitive and emotional biases that often encourage bad investment behavior, which in turn helps investors to stay on track towards their financial goals. 5 ways Robo advisors help prevent behavioral biases:

9.4.1 OBJECTIVE RISK ASSESSMENT

In terms of risk evaluation, Robo advisors combine mathematical tools to establish the amount of risk an investor could and should carry out. Robo advisors consider this, along with the timeline for investment, financial goals, and what we know about how much volatility equity markets have experienced in the past to provide a sense of how an investor can potentially withstand market dips as part of their portfolio. An objective assessment to counter biases—for example, risk aversion or prick sequences—leaves the investor with a more consistent investment strategy with his or her own risk profile.

9.4.2 GOAL-BASED INVESTING

Robo advisors: Goal-based investing. Robo-advisors introspect the needs of investors to plan their goals and time horizons. Robo advisors serve the purpose of keeping investors focused on long-term, goal-based investment outcomes like retirement planning, education financing, or wealth creation, and not short-term market volatilities or emotional outbursts. This goal-oriented approach helps deaden investors' perception of risk and loss by minimizing the effect of behavioral biases such as myopic loss aversion and recency bias.

9.4.3 DIVERSIFICATION AND ASSET ALLOCATION

Robo advisors promote diversified investment portfolios that are suitable to the risk appetite/aversion, as well as financial goals, of investors. Strategic beta: Robo advisors diversify trading throughout several asset classes, industries, and geographies, decrease performing focus risk, and help account longevity. Anchoring to an institutionally relevant asset allocation will immediately eliminate the influence of these biases, such as home bias (i.e., you love investing in your own country), or familiarity bias (meaning that you are likely to be overweight things that clutter through your day-to-day/week – sectors you know, etc.).

9.4.4 SYSTEMATIC REBALANCING

Growing portfolios are systematically rebalanced every couple of weeks to ensure investments stay within target allocation ranges. Furthermore, Robo-advisors do, in fact, rebalance their portfolios as market conditions change and the risk tolerance of their investors shift so that long-term goals are not compromised.

Disciplined rebalancing significantly reduces the tendency for portfolio drift and limits the detrimental effects of cognitive bias (e.g., anchoring or status quo bias) that otherwise may compel investors to "run winners too long" or avoid necessary changes to the investment mix.

9.4.5 BEHAVIORAL FINANCE INTEGRATION

Robo-advisors integrate lessons of behavioral finance into their presentation and communication to promote disciplined investing behavior. Robo-advisors offer nudges, reminders, and even educational elements to help inform investors of common behavioral biases and provide them with easy ways to counter common irrational decision-making tendencies. Robo-advisors reinforce more rational investor decisions by encouraging mindfulness and self-awareness, which help the investors to take less biased investment decisions.

Conclusively, Robo advisors play an important role in neutralizing the behavioral biases of investors and offer unbiased, evidence-based advice, which enables them to make rational and disciplined investment decisions.

Robo advisors improve investor empowerment, transparency, and long-term financial success in the digital age of wealth management with the help of technology and behavioral finance.

9.5 MAKING INVESTMENT DECISIONS IN THE DIGITAL ERA

Robo-advisers broke into the wealth-management field and heralded a new digital era in which investors have used technology to help them make the right investment decisions. Using sophisticated algorithms, data analytics, and psychological insights like behavioral finance, Robo-advisors help investors make those investment decisions with more confidence and in a way they can understand. Here are some of the forms in which Robo-advisers carve their contribution to investing in the digital era:

9.5.1 DATA-DRIVEN INSIGHTS

Robo advisors use big data and analytics to give investors access to market trends, asset performance, as well as portfolio optimization powered by big data and analytics. Robo advisors filter through enormous amounts of financial data in real-time, spotting where the investment opportunities and risks lie at that particular time. It enables investors to make informed decisions based on numbers and objective analysis, rather than human constitution and educated guesses. A data-driven approach also allows us to give investors more insight into what is driving performance in their portfolio and to make changes to their strategies accordingly.

9.5.2 TAILORED INVESTMENT STRATEGIES

Robo-advisors provide a customized investment strategy unique to each investor, aligned to their financial goals, risk tolerance, and investment horizon. Diversified portfolio building by Robo-advisors—a Robo-advisor constructs a varying assortment of investments that match the risk profile and goal-based investing framework for the investor in the data meant to be presented from the same. All-action that will not be made available by template-based solutions as well as course creation for anyone looking to succeed financially in the long run, making sure the offer fits personal goals and conditions.

9.5.3 ACCESSIBILITY AND CONVENIENCE

Robo advisors are revolutionary in terms of accessibility and convenience—literally 24/7 online, with web interfaces or a mobile app. With user-friendly dashboards and easy access to intuitive interfaces, investors can manage their portfolios, track returns, and receive live market news notifications from anywhere in the world. This provides investors with the data they need to make educated, proactive investment decisions in order to stay ahead of shifts in the market and strike while the iron is hot.

9.5.4 CONTINUOUS OPTIMIZATION

Robo-advisors are using complex machine learning (ML) algorithms, which allow them to tweak their diversified investment models in real-time based on the changing market. They help in capitalizing on market trends, investor behavior trends, candidates to increase the portfolio, and perhaps decrease some of the risks on a half-yearly basis, as well as asset allocations and rebalancing on behalf of the same. We take all this deep intelligence and aptly harness it via real-time monitoring of an ever-shifting constellation of inputs, thus enabling our dynamic investment management process to be no more static or sluggish than the market dynamics it was born to fluidly journey against over time, continually ensuring that long-term return potential remains abundant while also containing short-term risk to the downside.

9.5.5 BEHAVIORAL COACHING

They feature behavioral coaching, which is essentially a built-in component in their platform to assist investors in overcoming the usual biases we all have and emotions that can ruin our investing endeavor. Robo-advisors provide education and interactive tools along with personalized feedback to help investors navigate the bull markets and bear markets, keeping them focused on their long-term financial objectives to prevent emotional, panic-induced decision-making. This behavioral coaching will actually make the investors aware of themselves and their mental state so that in times when uncertainty prevails, they are likely to become wiser about investment rather than a scared investor.

From above, it becomes clear that Robo-advisors are the building blocks of transforming investment decision-making in the digital world, belonging with evidence-based thoughts, customized strategies, easy-to-access processes, and ongoing refinements. Behavioral coaching helps investors hold on to their financial objectives with higher support and transparency.

9.6 CONCLUSION

Robo advisors are a big leap for the wealth management industry, as they change the way we decide on our investments and financial planning opportunities in a world where everything is digital. This can be done through Robo advisors, which allow investors to manage their way through financial markets in a more efficient, transparent, and objective manner, depending on sophisticated technology and upon data used in behavioral finance.

In this chapter, we learned the in-depth layers of how Robo-advisors can actually save us from our behavioral biases, push us to invest all the better, and present the final value proposition: making every investor feel confident and crystal clear about his or her own investment objectives. From risk assessment and individual invest climate-based ML models to access, ongoing rebalancing, and behavioral counseling, Robo-advisors offer a full stack of functionalities to reinforce investment experiences and promote constructive habits that facilitate long-term financial health.

Additionally, unlike the traditional advisory model, Robo-advisors are zeroing in on ridding entry barriers and offering retail investors some of the same investment recommendations and portfolio management services through a professional-grade lens at a fraction of the cost, aiding in bringing equity to wealth management. It democratizes the wealth management process for greater financial inclusion and empowerment, thereby allowing every type of investor to accumulate wealth, achieve financial freedom, and realize their dreams.

Robo-advisors will be instrumental in shaping the future of wealth management. Make no mistake, the future of wealth management is going to be very digital, with Robo-advisors squarely at its heart, helping a generation that is not used to working with traditional banks. While it is unclear exactly what the FAs had in mind by saying "robos will never replace us," or to which brand of Robo-advice they were referring, there is no questioning that competition and expectations are going to decide the fate of the next generation of robos.

Nonetheless, it is important to note that Robo-advisors are not a cure-all for every issue investors may face and also do not erase the need for human intelligence and judgment in wealth management. Robo-advisors are great at automating the mundane, offering insight from data, and reigning in behavioral biases, but human advisors offer empathy, intuition, and tailored individual advice, especially when it comes to difficult or emotionally charged circumstances.

In the end, optimal wealth management solutions probably will draw on the strengths of both—melding Robo-advisor and human to support holistic experiences with a client-first perspective that reflects the varied preferred experiences of investors. Adopting this hybrid model and encouraging the cooperation of man and machine together can help forge new directions in innovation and value creation for the digital age within the wealth management industry.

KEYWORDS

- **behavioral biases**
- **digital transformation**
- **investment decisions**
- **robo-advisors**
- **wealth management**

REFERENCES

Abraham, F., Schmukler, S. L., & Tessada, J. (2019). Robo-advisors: Investing through machines. *World Bank Research and Policy Briefs* (134881).

Dziawgo, T. (2021). Wealth tech impact on wealth management sector.

Huang, R. H., Wang, C. C., & Zhang, O. X. (2022). The development and regulation of robo-advisors in Hong Kong: Empirical and comparative perspectives. *Journal of Corporate Law Studies, 22*(1), 229–263.

Jung, D., Glaser, F., & Köpplin, W. (2019). Robo-advisory: Opportunities and risks for the future of financial advisory. In *Advances in Consulting Research: Recent Findings and Practical Cases* (pp. 405–427).

Kowalewski, O., Pisany, P., & Ślązak, E. (2022). Digitalization and data, institutional quality and culture as drivers of technology-based credit providers. *Journal of Economics and Business, 121*, 106069.

Kumar, S., & Goyal, N. (2015). Behavioral biases in investment decision making–A systematic literature review. *Qualitative Research in Financial Markets, 7*(1), 88–108.

Quach, S., Thaichon, P., Martin, K. D., Weaven, S., & Palmatier, R. W. (2022). Digital technologies: Tensions in privacy and data. *Journal of the Academy of Marketing Science, 50*(6), 1299–1323.

Salo, A. (2017). Robo advisor, your reliable partner? Building a trustworthy digital investment management service (Master's thesis).

Silva, P. M. (2019). Robo-advising: Unfolding the risks. In *3rd International Conference on Computational Finance (ICCF2019)* (p. 214).

Singh, I., & Kaur, N. (2017). Wealth management through Robo advisory. *International Journal of Research-Granthaalayah, 5*(6), 33–43.

Uhl, M. W., & Rohner, P. (2018). Robo-advisors versus traditional investment advisors: An unequal game. *The Journal of Wealth Management, 21*(1), 44–50.

Ullah, S., & Elahi, M. A. (2014). Behavioral biases in investment decision making and moderating role of investor's type. In *SZABIST's 20th National Research Conference*, 10th May.

CHAPTER 10

Optimizing Investor Confidence: Robo-Advisory Utilizing Open Interest and Options Volume for Precise Large-Cap Stock Price Predictions

VIKAS SHARMA,* TEJINDER SINGH, and NITIN PATHAK

University School of Business, Chandigarh University, Mohali, India

*Corresponding author

ABSTRACT

This study interrogates the predictive efficacy of open interest and options volume in shaping large-cap stock price trajectories, employing robo-advisory mechanisms as analytical conduits. Anchored in the burgeoning landscape of financial technologies, the research examines how algorithmic advisory platforms synthesize derivative market indicators to generate precise, investor-centric insights. Utilizing a dataset of 28,088 observations from the National Stock Exchange (NSE) across 2021, the analysis incorporates Ordinary Least Squares regression to evaluate the explanatory capacity of open interest and options volume. Empirical findings substantiate the statistically significant role of both predictors, with open interest ($\beta = 0.54$) demonstrating superior predictive potency relative to options volume ($\beta = 0.27$). The adjusted R^2 value of 0.89 underscores the robustness of the model. Theoretically, this investigation enriches discourse on market microstructure by elucidating the informational content embedded within derivatives trading activity. Practically, it advocates robo-advisors as critical mediators capable of demystifying

Robo-Advisors and Artificial Intelligence in Human Resources Management: Revolutionizing HR.
Sanjay Taneja, Swati Gupta, Girish Lakhera, Mohit Kukreti, and Ercan Özen (Eds.)

DOI: 10.1201/9781779641410-10

esoteric option metrics for retail investors, thereby democratizing access to sophisticated forecasting tools. The research concludes that integrating open interest and options volume within robo-advisory frameworks not only enhances price discovery mechanisms but also fortifies investor confidence, contributing to market efficiency and inclusivity in emerging economies.

10.1 INTRODUCTION

Investment guidance is automated through algorithms powered by data analysis on factors like open interest and options volume, a service offered by Robo-advisors (Huang et al., 2022). Investors with limited knowledge of these metrics are served by these services, which offer insights into the significance of such metrics in predicting future prices (Khanna and Sharma, 2024). Simplifying complex indicators helps investors understand how open interest value and options volume value impact future pricing. By continually monitoring these dynamics, market opportunities aligned with investors' risk tolerance and goals are pinpointed by Robo-advisory systems, ensuring that lucrative prospects are not overlooked due to a lack of understanding. AI and Robo-advisory services have been leveraged by banks and NBFCs, including Paytm, to attract customers and bolster their market share (Singh, 2020; Singh, 2022). Furthermore, risk management is facilitated by these services through the recommendation of hedging strategies based on trend analyses, thereby mitigating potential losses from sudden market fluctuations. Personalized investment strategies tailored to an investor's preferences, risk appetite, and financial objectives are crafted by Robo-advisors, utilizing historical data and predictive analytics. Moreover, automated investment decisions are enabled through predefined algorithms, allowing opportunities to be capitalized on and portfolios managed efficiently with minimal manual intervention. In essence, Robo-advisory services serve as a valuable bridge for investors seeking clarity on open interest and options volume. They offer actionable insights, facilitate decision-making, and provide personalized strategies, ultimately empowering investors to leverage these metrics effectively in predicting future prices. Stock options form a part of the futures and options segment. These options are derived from the underlying securities. The derived data is a very important part of the overall system performance (Wexler and Oberlander, 2022).

The influence of option markets permeates throughout the underlying stock prices (Roszkowska, 2021). This chapter focuses only on the options segment, with a selection of stock options as a sample. Stock options are of two types: call and put. The buyer buys a call option if he is bullish on the stock. Similarly, a put option is bought by the seller if he is bearish on the stock. Premium is paid by the buyers of these options, which is calculated by multiplying the prevailing rate with the number of shares in one market lot. Each share has a different number of shares allotted in one market lot as per the market capitalization and as decided by the exchange at a central level under the regulation of the market regulator.

Open interest denotes the unclosed positions of options contracts with stock and index as underlying assets within a specified expiry month on a given day. Historically, the correlation between open interest and trading volume has been a subject of significant interest, particularly during merger announcements. Analysts attempted to predict individual stock prices by analyzing the distribution of open interest in associated options contracts (Bhuyan and Chaudhury, 2005). The NSE has been considered because, from the meager numbers in the initial stage, it has emerged as the world's number one derivative exchange, where options are the major traded segment. The introduction of options in NSE was a turning point in the history of Indian capital markets. The year 2001 was significant since it was the year when the market first accepted index options trading as well as single stock futures and options trading. Currency options trading started in 2009. NSE held the top spot as the world's largest derivatives exchange for three consecutive years, from 2019 to 2021. It is crucial to study and analyze the market for options and its data accuracy in order to accurately detect the potential impact on future prices. The role of open interest options can also be gauged with the help of the Put–Call Ratio. This ratio is commonly used in predicting market returns, whether in a bullish or bearish market (Jena et al., 2019). Past studies have demonstrated the significant impact of open interest with respect to changes in futures prices, particularly in the commodity futures segment. A higher number of contracts is associated with a decrease in commodity futures prices, according to the study. During volatile market movements, volume results are more positive than open interest (Dungore and Patel, 2021). Analyzing open interest and trading volume provides valuable insights for market participants in managing prices. Furthermore, collectively observing the trend of changes in the prices, number of contracts, and values of open interest makes it easier to track the purchasing or selling impetus driving market fluctuations (Szczygielski et al., 2023). However, some studies believe that put options are used for speculation, whereas call options are used for the purpose of hedging trading (Jongadsayakul, 2019). Compared to individual stock options, hedgers show a stronger preference for the index options market (Daud et al., 2022). It turns out that the liquidity of the options market for a large stock index is linked to the liquidity of funds (Basha et al., 2021). Since more and more uninformed or irrational traders and rational or informed traders are attracted to futures trading, stock return volatility increases (Robbani and Bhuyan, 2016). Some studies observed an abnormal open interest increase immediately before the large price decline (Sharma et al., 2023). When used collectively, VWAP, moving average, open interest, and volumes specified for a given limit help assess current market stock for decision-making in the intra-day trading in the options market (Maurya et al., 2021). The remaining part of this research article is organized in the following manner: It discusses the theoretical framework of the options. Subsequently, it consists of a literature review, formulation of predictors and their equations, research methodology, result analysis, conclusion, limitation, and scope for further research (Sharma et al., 2024). The trends in open interest and options volumes for large-cap stocks are given below in chart form for the period under study. Open interest has grown over the previous month's figures in all months except for four. Likewise, options volume has increased over the preceding month's figures, barring three months. The average monthly open interest for large caps during 2021 was 120,078, while the figure for options volume stood at 193,837 (refer to Figure 10.1). To trace the history of introducing options, the U.S. options exchange (CBOE) pioneered options in 1973, becoming the world's first exchange to do so. In India, stock options were introduced in 2002, initially limited to a select number of stocks.

However, by 2021, the derivatives segment witnessed the trading of more than 150 stocks, as per information available from the NSE.

It is evident from Table 10.1 that an investor can buy call or sell put for bullish positions and sell call or buy put for bearish positions.

It is evident from Table 10.2 that an investor may have different behaviors: bullish, bearish, unwinding, and short covering. These behaviors depend upon an increase or decrease in option prices and option open interest. The accuracy of the Black-Scholes model (BSM) holds good even

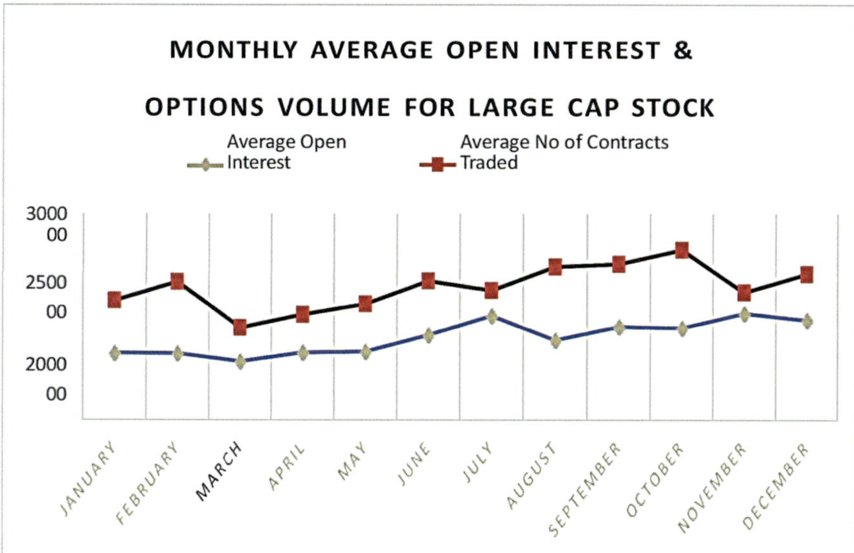

FIGURE 10.1 Average options open interest and volume during 2021.
Source: By the authors.

TABLE 10.1 Stock Option Types: Call and Put (Investor Expectation)

BUY CALL (Bullish)	BUY PUT (Bearish)
SELL CALL (Bearish)	SELL PUT (Bullish)

Source: Author's work.

TABLE 10.2 Investor Behaviors

Option Prices	Option Open Interest	Likely Behavior of the Investors or Traders
Increase	Increase	They are bullish
Decrease	Decrease	They are unwinding their long-term position
Decrease	Increase	They are bearish
Increase	Decrease	They are short-covering their position

Source: Author's work.

in the current economic and financial environment. Our analysis found no statistically significant difference between observed market prices and the BSM model's theoretical estimates for option prices. The ANOVA technique verified the results (Bhatia et al., 2021). Nonetheless, certain investigations reveal that amidst the subprime and COVID crises, the efficacy of these equations markedly diminishes (Redroban and Cifuentes, 2021). A survey of the German markets reveals that clients of discount broking houses prefer to invest for purely speculative purposes along with gambling (Oehler and Schneider, 2022). With the decrease in liquidity, mispricing window size also increases using the intraday market data (Shankar et al., 2019). Stock options, especially call options, reflect recent information compared to the underlying stock prices. Unusual trading takes place in the index option contracts before major price shocks for many events (Spyrou, 2011). Pathak and Rastogi (2010) used the combined open interest and volume predictors and regressed them against the future prices of the near-month expiry. Since most studies are either on the interlinkages between the spot market and options or the impact of corporate events on different markets, there has been very limited research in the case of the independent variables called open interest and volume predictors.

10.2 REVIEW OF LITERATURE

Our investigation is driven by the identification of significant lacunae in the current body of research pertaining to the equity derivatives sector. Predominantly, prevailing studies originate from Western jurisdictions and are characterized by relatively brief temporal scopes. Surprisingly, our review found no major studies that exclusively examined large-cap stocks. The researchers, Jiang et al. (2019) scrutinized the behavior of investors in the U.S. markets concerning their engagement with options when in possession of adverse information. Their analysis revealed that options are prominently and efficiently employed to capitalize on negative information. Our data indicates that the options market's difference between a stock's actual trading price and its expected price reached its maximum point two weeks prior to the height of short sell trades. The options market plays a crucial role in expanding investment opportunities by mitigating limitations imposed by arbitrage constraints in the traditional stock market. The continued participation of informed investors in this market underscores its reputation as a reliable platform, without compromising overall performance of the markets. Pathak et al. (2015) conducted an examination of the informational implications within equity index options trading using one of the largest exchange contracts traded on a major stock exchange in India.

Methodological approaches encompassed the utilization of linear regression to explore relationships and the auto-regression model to scrutinize market information content. Findings from the VAR analysis indicate a consistent preference among informed traders for at-the-money options contracts within the Indian market throughout the studied periods. However, notable observations reveal a significant leadership role of the spot market for the non-at-the-money options classes, transcending various trends in the market. Additionally, the study underscores the noteworthy leadership of the spot market across different market trends and options classes. In simpler terms, the study by Chakravarty et al. (2004) investigated how option markets influence stock prices. The researchers looked at data on stock prices and the prices implied by options to see which market played a bigger role in setting the stock price. They found that the importance of option markets varied depending on the specific stock. Options seemed to be more informative when there was a lot of trading in options contracts, not a lot of trading in the stock itself, narrow gaps between the prices at which options could be bought and sold, and large gaps between the prices at which the stock could be bought and sold. Wasiuzzaman (2022), in their examination of the Tadawul All-Shares Index, observed that higher trading volume might signify a surge in investor sentiment. Agarwalia Sobhesh Kumar (2013) investigated the impact of expiration day on intraday volatility, noting a discernible increase in stock volatility over the course of several hours on expiration days, which is a noticeable difference compared to non-expiration days. In one of the studies, researchers found (Jangir et al., 2023) that the volatility dynamics within the spot market were influenced by the "expiration day effect," stemming primarily from the actions of arbitrageurs and speculators. Empirical evidence documented that this effect was notably influenced by the settlement procedures adopted within a given market. Dungore and Patel (2021) conducted a study to determine the existence of a causal relationship between fluctuations, turnover, and open interest for derivative contracts sold on a major market platform, as well as to ascertain the magnitude and direction of these relationships. The research found that open interest value had less of an impact on volatility than volume. Additionally, it elucidates the interrelation among fluctuations, turnover, and open interest. The findings reveal that the majority of volatility variance stems from itself. Moreover, the effect of volume on volatility surpasses that of open interest. The estimation of Granger causality indicated the presence of noise trading, as evidenced by the existence of unidirectional causality in only two instances. The study investigated the dynamic directional trading within stock limit order books and options markets, employing a sample of NYSE-listed stocks. The findings revealed a mutual causality between stock limit order book aggressiveness and directional trading in the options markets, as demonstrated by Granger causality analysis (Costello et al., 2020). Badgi (2022) investigated the potential of non-price data, particularly changes in Put-Call Ratio (PCR) Open Interest (OI), volume, and volatility, to enhance market price forecasting accuracy. The research employed the open interest band of the PCR to predict Nifty price movements. The findings suggest that fluctuations in PCR Open Interest levels demonstrate a significant correlation with subsequent Nifty returns. Conversely, volume levels exhibited comparatively weaker correlations and predictability. Hence, alterations in volume levels were deemed insufficient for predicting Nifty returns. This study suggests options are stacked in our favor. With high open interest, there's a strong bet the Nifty will stay below the band average. This creates a juicy opportunity for call writers like us. We can sell calls and collect premium, banking on the market staying put or even dipping. It's basically a chance to earn easy money if the Nifty behaves as expected. Just remember, this is a gamble, and unexpected price movements can be loss making if investors are not careful. Consequently, market makers are inclined to bet that the option will expire out-of-the-money (OTM), enabling the seller to attain maximum profit. Ryu and Yang (2018) conducted an examination into the directional informational content of options volumes and their capacity to forecast returns. This study examined the link between put-call volume ratios and decoupled volume ratios, utilizing a detailed, high-frequency dataset that captured investor and trade specifics. The findings revealed that foreign investment firms have been identified as active traders whose options transactions strongly predict the movements of underlying spot returns. Particularly, their open-buy trades demonstrated a significant ability to predict next-day spot returns, albeit with diminishing predictive efficacy over extended time horizons, indicative of their transient superiority in short-term directional information within index options trading. A new study shows foreign investors are getting good at predicting short-term stock price movements. They do this by buying options contracts that reflect their belief a stock will go up or down in price soon. These options have short expiration times. The study found the bigger the investment and the riskier the option (meaning it's less likely to pay off unless the stock price moves significantly), the better these investors are at predicting those price moves. This strategy works even better in markets where investors can't easily bet on stocks going down by borrowing and selling them first (short selling). In simpler terms, by looking at which options foreign investors are buying, especially short-term options that are unlikely to be profitable unless the stock price moves significantly, and focusing on markets with restrictions on short-selling, we can get clues about how the stock price might move in the short term (DeLisle et al., 2016). This research by Li (2021) reveals how options trading can influence companies, especially those struggling. Companies with limited information or trouble getting funding seem to be more impacted by options activity. This suggests options traders might be acting like information brokers. Their trades might influence the decisions and policies of these companies. In short, options trading can affect how some companies, particularly those facing difficulties, are run. This study by Martín et al. (2016) examined at how options trading behaved around the time companies release their earnings reports. They found that there was a lot more options trading happening in the lead-up to and on the day of the announcement itself. However, a key distinction emerged: informed traders seemed to focus on opening new positions exclusively in non-in-the-money (ITM) options. In contrast, the increase in trading activity for ITM options was due to traders closing out existing positions. This suggests that if informed traders are using their knowledge to make profits, they likely focus on ATM and OTM options. Interestingly, the study also found that even with low activity (a "thin" market), having an options market available improves the pricing accuracy in the stock market. Avinash and Mallikarjunappa (2020) findings corroborate existing literature which discovered similar evidence favoring open interest-based price predictors exclusively. The results obtained from using a regime-switching model, along with confirmatory evaluating the fit within the sample and forecasting performance outside the sample, support the hypothesis that non-price option variables, specifically open interest and volume, contain valuable information predictive of the underlying asset's price at option contract expiry. Moreover, this suggests that the price derived from open interest and volume weights, modeled using MIDAS methodology, could assist market participants, particularly those with limited information, in anticipating future spot prices on options' maturity day (Seo and Wachter, 2019). This study investigates whether informed traders and hedgers in the options market can leverage changes in open interest to predict future stock price movements. The underlying premise is that informed traders possess superior knowledge of future economic conditions, enabling them to anticipate price trends ahead of uninformed investors. The study discerned distinctions between the information imparted by options open interest and that derived from options trading volume. Specifically, it affirmed that the increase in options open interest offers unique insights into future stock market returns, distinct from conventional variables such as options trading volume. The research by Lubnau and Todorova (2012) explored the feasibility of integrating option open interest into effective chart-based trading strategies. A comprehensive compilation of more than 2000 trading strategies was employed, targeting the leading Index in Germany as well as the large cap shares listed on the German exchange by market capitalization. By incorporating open interest data alongside traditional spot market indicators, the accuracy of technical trading rules can be significantly improved. Notably, both put and call open interest demonstrate potential in providing insights into future equity prices, whereas the open interest differential exhibited notably poor performance (Gupta et al., 2023). Studies investigating the predictive power of options for underlying asset prices highlight the potential of specific option types. Prior research indicates that OTM calls and ITM puts demonstrate particularly strong predictive capabilities. Interestingly, a separate study investigating the co-movement between volatility, trading volume, and open interest in the futures and options segment reported a more pronounced influence of volume for volatility with reference to open interest. This finding suggests the potential presence of noise trading in such circumstances. These findings translate into practical applications. Actively trading based on open interest data can potentially generate higher returns than passive strategies, while equity options open interest information proves valuable not only for speculative trading, but also serves as a valuable tool for risk management in hedging strategies (Bhuyan and Chaudhury, 2005). Research indicates that the options market holds particular significance for informed traders leading up to major events. This is important for academics to consider, as these informed traders can potentially cause significant shifts in wealth distribution among investors. Consequently, regulators should focus additional resources on monitoring options markets to proactively identify and prevent potential insider trading activity (Liu et al., 2015). This study examines the persistence of the number of contracts of open interest and option volume in influencing future prices, aiming to verify consistency with previous research (Sharma et al., 2024).

Formulation of Predictors and Regression Equation

All the stock options have a different set of strike prices for calls and puts for near, next, and far-away expiry. The predictors were obtained using Bhuyan and Williams's methodology (2004). It is assumed that the strike prices for calls are as follows:

$$\{X_c^i, i = 1, 2, 3\ldots\ \ldots, k\}$$

likewise, strike prices for put options for t are taken as:

$$\{X_l^P, l = 1, 2, 3\ldots\ \ldots.\ , L\}$$

The values of open interest in case of calls for strike prices for t day is denoted by O^c_{it}

The values of open interest in case of put options strike prices for t day is denoted by O^P_{it}

Afterwards, value of open interest in case of call options is found as follows:

$$\Sigma_{i=1}^{K} X_i^C \text{where } w_{it}^C = \frac{O_{it}^c}{\Sigma_{i=1}^{K} O_{it}^c}$$

The open interest predictor for put options is calculated as follows:

$$O_t^P = \Sigma_{l=1}^{L} X_l^P w_{it}^P \text{ where } w_{it}^P = \frac{O_{it}^P}{\Sigma_{l=1}^{L} O_{it}^P}$$

w^C_{it} and w^P_{lt} displays the weighted values of open interests attached to the strike prices of call options and put options. To calculate the value of options volume predictor for various strike prices, value of the volume-based call predictor is determined using the following formula:

$$V_t^C = \Sigma_{i=1}^{K} X_i^C q_{it}^C \text{where } q_{it}^C = \frac{O_{it}^C}{\Sigma_{i=1}^{K} O_{it}^C}$$

The value of volume-based put predictor is calculated using following method:

$$V_t^P = \sum_{l=1}^{L} X_i^P q_{it}^P \text{ where } q_{it}^P = \frac{V_{it}^P}{\sum_{i=1}^{K} O_{it}^P}$$

Price discovery results from the interaction of demand and supply. In a similar fashion, both call and put options are integrated to create a Combined Value of Open Interest Predictor and a Combined Value of Volume-based Predictor. These values are found as follows:

$$\text{Combined Open Interest Predictor} = \frac{\sum_{i=1}^{K} o_t^c O_{it}^c}{\sum_{i=1}^{K} O_{it}^c} + \frac{\sum_{l=1}^{L} o_t^p O_{it}^p}{\sum_{l=1}^{L} O_{it}^p}$$

$$\text{Combined Volume based Predictor} = \frac{\sum_{i=1}^{K} v_t^c V_{it}^c}{\sum_{i=1}^{K} V_{it}^c} + \frac{\sum_{l=1}^{L} o_t^p O_{it}^p}{\sum_{l=1}^{L} O_{lt}^p}$$

Where Θ_t^C and o_t^P signify values of the open interest predictors, v_t^c and v_t^P signicy values of the volume-based predictors, and v_{it}^c and v_{it}^p Denote volumes at different strike prices on sample trading days.

The Combined Open Interest Predictor (COP) and Combined Volume-based Predictor (CVP) are employed to forecast future prices at maturity T for near-month expirations. COP and CVP are calculated over three sample trading days and subsequently regressed against the future maturity prices to evaluate their predictive capacity for price changes at maturity, specifically for near-month expirations.

10.3 METHODOLOGY AND ANALYSIS

The study identifies the top 10 large-cap stocks based on market capitalization on the National Stock Exchange (NSE) for the year 2021. The NSE is chosen for this analysis because it is the biggest derivatives exchange in the world in terms of volume for the third year in a row in 2021.

Sample Size: Table 10.3 shows that only stock options have been considered for the study. The total number of observations initially amounted to 61,000 for the study. After excluding non-exercised strike prices characterized by zero open interest and zero volume, the net number of observations decreased to 28,088. On average, each stock was represented by approximately 2,500 observations, except for two stocks, which had a higher number of observations exceeding 4,000 each.

Future maturity prices are the dependent variable. The following are the predictors: (1). The open interest in stock options; and (2). The volume of stock options. The statistical package employed for the analysis was EViews. The statistical technique applied was ordinary least squares (OLS). Monthly data pertaining to open interest and option volumes, measured in terms of contract numbers, were computed for stock options across three non-expiry days. One time segment is one month. 12-time segments were taken during the period of the study.

TABLE 10.3 List of Large Cap Stocks

SL. No	Large Cap Stocks (2021)
1	RIL
2	TCS
3	The HDFC Bank
4	Infosys Ltd
5	HUL
6	HDFC Ltd
7	ICICI
8	KBL
9	SBI
10	BAJ FIN

Source: National Stock Exchange.

The following hypotheses were tested: H0: The open interest does not serve as a predictor of future prices. H1: The open interest serves as a predictor of future prices. H0: Future prices are not predicted by the options volume. H1: Future prices are predicted by options volume. The study is set to span from January 2021 through December 2021.

10.4 FINDINGS

In this investigation, the influence of two independent variables, the value of the joint Open Interest Predictor and joint Options Volume (measured in contracts), on monthly future prices nearing expiry was assessed through regression analysis. To ensure the validity and reliability of the statistical results, normality of the data was achieved by applying a logarithmic transformation to all variables using Excel software. Furthermore, stationarity of the data was rigorously confirmed through the implementation of an Augmented Dickey-Fuller (ADF) test conducted within EViews software. The subsequent passage will detail the hypothesis tested.

$$H0{:} = 1(\text{Nonstationary}),\ H1{:} < 1(\text{Stationary})$$

The volume was found to be stationary at a level because the p-value was less than 0.05. Hence, the null hypothesis that volume has a unit root is rejected. The level of open interest was also found to be stationary when the p-value was less than 0.05. Hence, the null hypothesis that open interest has a unit root is rejected. Monthly future prices were found to be non-stationary at a first level since the p-value was greater than 0.05. Hence, the null hypothesis that open interest has a unit root is accepted. Therefore, monthly future prices were observed to be stationary using ADF at the first difference.

The following regression equation was then used for applying the pooled OLS regression in EViews: Pt= c + OI + V, where Pt (dependent variable) denotes the values of future prices at maturity in the case of near-month expiry, O.I. is the value of open interest, and the predictor V is the volume and independent variable.

TABLE 10.4 Results of the Regression of Open Interest and Volume Against Future Prices

Input	Coefficient	Std. Error	t-Statistic	P-Value
Fixed	4.181	0.087	47.968	0.00
OI Value	0.545	0.056	9.643	0.00
Contracts	0.277	0.061	4.478	0.00
R-Squared	0.898			
Adj R- Squared	0.898			

The coefficient of open interest is 0.54, volume is 0.28, and the total adjusted R-squared is 0.89. The model achieved statistical significance in all cases, with the open interest predictor coefficients also being significant across all significance levels. The R2 of the regression was observed as 0.89, indicating high predictive ability. As per the observations from

Table 10.4, the results of the hypothesis are discussed as follows: The analysis resulted in the rejection of the null hypothesis that open interest has no predictive power on future prices. This conclusion is supported by a p-value less than 0.05, implying a significant influence of open interest on future price prediction. Similarly, the null hypothesis that option volumes lack predictive power for future prices was also rejected due to a p-value less than 0.05. This suggests that option volumes also exert a noteworthy impact on forecasting future prices. There is control over the standard error, and the findings may be relied upon. The open interest is comparatively more predictive, with a 0.54 coefficient, as compared to the volumes, which have a coefficient of 0.27. As per the existing work of Rajesh Pathak (2010), the volumes were insignificant when they were combined with open interest. However, the volumes were significant when their impact was checked on future prices in isolation (Table 10.4).

10.5 RESEARCH IMPLICATIONS

Due to the emerging phase of financial markets and an underdeveloped options market in the case of the NSE, the predictiveness of open interest and options volume may not be suitably reflected (Pathak and Rastogi, 2010). In this chapter, the results are different from previous studies in the sense that volumes, when used with open interest, are also significant. This was possibly due to the one-year duration as compared to the short sample duration of 3 months, taken as the duration of the study by Rajesh Pathak (2010). To ensure that multicollinearity exists, the variance inflation factor (VIF) was checked among the independent variables, which was 7. The VIF is reasonable at 7.0, indicating that multicollinearity is not severe among the predictors (Sharma et al., 2023).

10.6 CONCLUSION

The findings of this chapter, post-hypothesis testing, establish the consistent predictive ability of both open interest and option volumes regarding future prices. Despite India's large pool of retail investors and its dominant position in derivatives, many people are left out due to the market's complexity. A new study found that open interest data, which is a better predictor of future prices than volume, is a significant barrier for average investors. These notions are harder to grasp than basic price movements.

The report suggests Robo-advisors as a solution. These services can translate complex data and inform users about how it influences prices. Furthermore, regulators should think about creating retail-friendly technologies that convert this data into useful insights regarding pricing changes. By promoting user-friendly products and enhancing investor education, more people will be able to securely participate in the derivatives market, benefiting both themselves and the market overall. This, in turn, can promote competition within an established regulatory framework and help create a more stable financial climate by diverting regular investors from the risky world of cryptocurrencies.

10.7 LIMITATIONS OF THE STUDY AND SCOPE FOR FUTURE RESEARCH

This study specifically examines stock options that have a significant market capitalization. Subsequent studies could investigate the relevance of these discoveries to mid-cap and small-cap equities and indices, examining potential variations in their correlation with market capitalization. The period of the current chapter is one year, as a result of the extensive amount of observations. The time duration may also be extended for further studies.

KEYWORDS

- **large-cap stocks**
- **open interest**
- **options volume**
- **price prediction**
- **robo-advisory**

REFERENCES

Agarwalia, S. K., & P., A. (2013). Expiration-day effects and the impact of short trading breaks on intraday volatility: Evidence from the Indian market. *Journal of Futures Markets, 33*(11), 1046–1070.

Avinash, & Mallikarjunappa. (2020). Informational role of open interest and transaction volume of options: A meta-analytic review. *FIIB Business Review, 9*(4), 275–285. https://doi.org/10.1177/2319714520980662 (accessed on 11th August 2025).

Badgi, V. (2022). Open interest band – as a predictor of underlying future price: An empirical evidence with respect to option market in India. *03*, 137–157. https://doi.org/10.17605/OSF.IO/Z5JCF (accessed on 11th August 2025).

Basha, S. A., Elgammal, M. M., & Abuzayed, B. M. (2021). Online peer-to-peer lending: A review of the literature. *Electronic Commerce Research and Applications, 48*. https://doi.org/10.1016/j.elerap.2021.101069 (accessed on 11th August 2025).

Bhatia, A., Chandani, A., Atiq, R., Mehta, M., & Divekar, R. (2021). Artificial intelligence in financial services: A qualitative research to discover robo-advisory services. *Qualitative Research in Financial Markets, 13*(5), 632–654. https://doi.org/10.1108/QRFM-10-2020-0199 (accessed on 11th August 2025).

Bhuyan, R., & Chaudhury, M. (2005). Trading on the information content of open interest: Evidence from the US equity options market. *Derivatives Use, Trading & Regulation, 11*(1), 16–36. https://doi.org/10.1057/palgrave.dutr.1840004 (accessed on 11th August 2025).

Chakravarty, S., Gulen, H., & Mayhew, S. (2004). Informed trading in stock and option markets. *Journal of Finance, 59*(3), 1235–1258. https://doi.org/10.1111/j.1540-6261.2004.00661.x (accessed on 11th August 2025).

Costello, A. M., Down, A. K., & Mehta, M. N. (2020). Machine + man: A field experiment on the role of discretion in augmenting AI-based lending models. *Journal of Accounting and Economics, 70*(2–3). https://doi.org/10.1016/j.jacceco.2020.101360 (accessed on 11th August 2025).

Daud, S. N. M., Ahmad, A. H., Khalid, A., & Azman-Saini, W. N. W. (2022). FinTech and financial stability: Threat or opportunity? *Finance Research Letters, 47*. https://doi.org/10.1016/j.frl.2021.102667 (accessed on 11th August 2025).

DeLisle, R. J., Lee, B. S., & Mauck, N. (2016). The dynamic relation between options trading, short selling, and aggregate stock returns. *Review of Quantitative Finance and Accounting, 47*(3), 645–671. https://doi.org/10.1007/s11156-015-0516-2 (accessed on 11th August 2025).

Dungore, P. P., & Patel, S. H. (2021). Analysis of volatility volume and open interest for Nifty index futures using GARCH analysis and VAR model. *International Journal of Financial Studies, 9*(1), 1–11. https://doi.org/10.3390/ijfs9010007 (accessed on 11th August 2025).

Gupta, M., Taneja, S., Sharma, V., Singh, A., Rupeika-Apoga, R., & Jangir, K. (2023). Does previous experience with the Unified Payments Interface (UPI) affect the usage of central bank digital currency (CBDC)? *Journal of Risk and Financial Management, 16*(6), 286.

Huang, R. H., Wang, C. C., & Zhang, O. X. (2022). The development and regulation of robo-advisors in Hong Kong: Empirical and comparative perspectives. *Journal of Corporate Law Studies*. https://doi.org/10.1080/14735970.2021.2012884 (accessed on 11th August 2025).

Jena, S. K., Tiwari, A. K., & Mitra, A. (2019). Put–call ratio volume vs. open interest in predicting market return: A frequency domain rolling causality analysis. *Economies, 7*(1), 1–10. https://doi.org/10.3390/economies7010024 (accessed on 11th August 2025).

Jiang, Z., Kang, S. H., Cheong, C., & Yoon, S. M. (2019). The effects of extreme weather conditions on Hong Kong and Shenzhen stock market returns. *International Journal of Financial Studies, 7*(4). https://doi.org/10.3390/ijfs7040070 (accessed on 11th August 2025).

Jongadsayakul, W. (2019). Determinants of investor behavior in SET50 index futures and options markets: Evidence from Thailand futures exchange. *Journal of International Studies, 12*(3), 22–30. https://doi.org/10.14254/2071-8330.2019/12-3/2 (accessed on 11th August 2025).

Khanna, K., & Sharma, V. (2024). Business decision-making with an analytical approach: A new leadership pattern. In *Global Financial Analytics and Business Forecasting*. https://www.scopus.com/inward/record.uri?eid=2-s2.0-85184428077&partnerID=40&md5=f70a811294cc3df9d00cfd4cf25735f3 (accessed on 11th August 2025).

Lubnau, T. M., & Todorova, N. (2012). Technical trading with open interest: Evidence from the German market. *Applied Financial Economics, 22*(10), 791–809. https://doi.org/10.1080/09603107.2011.627210 (accessed on 11th August 2025).

Martín, J. G., Piqueras, B. H., & Escribano, A. M. I. (2016). The informational role of thin options markets: Empirical evidence from the Spanish case. *Estudios de Economia, 43*(2), 233–263. https://doi.org/10.4067/s0718-52862016000200004 (accessed on 11th August 2025).

Maurya, R. K., Rai, P., & Yadav, H. B. (2021). Decision making in intraday index options trading: A parametric approach. In *Proceedings of the Confluence 2021: 11th International Conference on Cloud Computing, Data Science and Engineering* (pp. 1067–1071). https://doi.org/10.1109/Confluence51648.2021.9377182 (accessed on 11th August 2025).

Oehler, A., & Schneider, J. (2022). Gambling with lottery stocks? *Journal of Asset Management, 17*, 700156756, 41260. https://doi.org/10.1057/s41260-022-00268-y (accessed on 11th August 2025).

Pathak, R., & Rastogi, N. (2010). Informational role of options opens interests and volume in forecasting future prices: A study on Indian market. *The IUP Journal of Financial Economics, VIII*(3), 49–61.

Redroban, S., & Cifuentes, A. (2021). On the performance of the Black and Scholes options pricing formulas during the subprime and COVID-19 crises. *Journal of Corporate Accounting & Finance, 32*(3), 75–85. https://doi.org/10.1002/jcaf.22504 (accessed on 11th August 2025).

Robbani, M. G., & Bhuyan, R. (2016). Options on a stock index: Volume and volatility–Empirical. pp. 187–188.

Roszkowska, P. (2021). FinTech in financial reporting and audit for fraud prevention and safeguarding equity investments. *Journal of Accounting and Organizational Change, 17*(2), 164–196. https://doi.org/10.1108/JAOC-09-2019-0098 (accessed on 11th August 2025).

Ryu, D., & Yang, H. (2018). The directional information content of options volumes. *Journal of Futures Markets, 38*(12), 1533–1548. https://doi.org/10.1002/fut.21960 (accessed on 11th August 2025).

Seo, S. B., & Wachter, J. A. (2019). Option prices in a model with stochastic disaster risk. *Management Science, 65*(8), 3449–3469. https://doi.org/10.1287/mnsc.2017.2978 (accessed on 11th August 2025).

Sharma, V., Jangir, K., Gupta, M., & Rupeika-Apoga, R. (2024). Does service quality matter in FinTech payment services? An integrated SERVQUAL and TAM approach. *International Journal of Information Management Data Insights, 4*(2), 100252. https://doi.org/10.1016/j.ijimei.2024.100252 (accessed on 11th August 2025).

Sharma, V., Taneja, S., Gupta, M., & Ozen, E. (2023). Impact of service quality on behavioral intention to use FinTech payment services: An extension of SERVQUAL model. *Asia Pacific Journal of Information Systems, 33*(4), 1093–1117.

Sharma, V., Taneja, S., Jangir, K., & Khanna, K. (2024). Green finance: An integral pathway to achieving sustainable development. In *Sustainable Investments in Green Finance* (pp. 49–63). IGI Global.

Singh, T. (2020). FinTech adoption: A critical appraisal of Paytm in India. *The IUP Journal of Management Research, XIX*(4), 7–17.

Singh, T. (2022). Artificial intelligence–Application in the field of Indian banking sector. *Artificial Intelligence–Application in the Field of Indian Banking Sector, 020002*(October).

Szczygielski, J. J., Charteris, A., & Obojska, L. (2023). Do commodity markets catch a cold from stock markets? Modelling uncertainty spillovers using Google search trends and wavelet coherence. *International Review of Financial Analysis, 87*, 102304.

Wasiuzzaman, S. (2022). Impact of COVID-19 on the Saudi stock market: Analysis of return, volatility and trading volume. *Journal of Asset Management, 23*(4), 350–363. https://doi.org/10.1057/s41260-022-00269-x (accessed on 11th August 2025).

Wexler, M. N., & Oberlander, J. (2022). Robo-advice (RA): Implications for the sociology of the professions. *International Journal of Sociology and Social Policy*. https://doi.org/10.1108/IJSSP-09-2021-0245 (accessed on 11th August 2025).

CHAPTER 11

The Rise of Robo-Advisors and Shifting Consumer Purchase Intentions

SNEHA SINGH,[1,*] DEEPAK KAUSHAL,[1] and LARISA MISTREAN[2]

[1]*Department of Management Studies, Graphic Era Deemed to be University, Dehradun, Uttarakhand, India*

[2]*Department of Investments and Banking Activity, Faculty of Finance, Academy of Economic Studies of Moldova, Chisinau, Moldova*

*Corresponding author

ABSTRACT

The ascendancy of robo-advisors epitomizes a transformative epoch in financial services, where automation, artificial intelligence, and data analytics converge to redefine consumer purchase intentions and wealth management paradigms. This study investigates the interplay between robo-advisory platforms and consumer decision-making, highlighting how cost-efficiency, accessibility, personalization, transparency, and trust collectively stimulate adoption. Robo-advisors transcend traditional advisory limitations by offering scalable, real-time, and inclusive financial guidance, thereby democratizing access to investment strategies once reserved for high-net-worth individuals. Leveraging machine learning algorithms, these platforms generate individualized, goal-based recommendations, fostering consumer confidence through data-driven precision and algorithmic transparency. Yet, challenges such as the absence of human empathy, algorithmic biases, regulatory complexities, and cybersecurity vulnerabilities remain critical impediments to widespread acceptance. Opportunities, however, lie in hybrid advisory models, ESG-oriented

Robo-Advisors and Artificial Intelligence in Human Resources Management: Revolutionizing HR. Sanjay Taneja, Swati Gupta, Girish Lakhera, Mohit Kukreti, and Ercan Özen (Eds.)

DOI: 10.1201/9781779641410-11

investment strategies, and advanced educational initiatives that enhance financial literacy while bridging digital divides. By aligning technological innovation with consumer-centric values, robo-advisors are poised to serve not merely as disruptive forces but as enablers of sustainable, ethical, and inclusive financial ecosystems. The findings underscore that the future of robo-advisory adoption depends on balancing algorithmic efficiency with human sensibility, thereby reshaping both consumer intentions and the broader architecture of global financial intermediation.

11.1 INTRODUCTION

The use of Robo-advisors has been increasingly gaining importance within the volatile realm of financial services, which is conventionally known to be infallible. In its more evolved expositions, Robo-advisors have not only proven to be potential challengers to incumbents, but have also revolutionized modern-day wealth management, thereby resetting consumers' perceptions about investment choices (Singh et al., 2019). This chapter begins a complex journey to the heart of this financial revolution, unwinding the different layers that make up the rise of Robo-advisors and their delicate waltz with consumer purchase intentions (Song and Kim, 2022). This leads to a bigger market for Robo-advisory services and us being, yet again, re-monetized. In a world dominated by digital technology, most things are moving towards being done automatically—online shopping, online cab services, digital transactions, online banking platforms, so forth, (Singh and Kaur, 2017). In simpler terms, we can say that automation has become an inseparable part of our present lives. Every corporate function has seen automation but is now starting to see the effects of automation trickling down into financial decision making (Abraham et al., 2019). Decision-making is a human task by its very nature. Decision-making is simply choosing the best way out of all possible ways. Digital financial advisors, the offspring of artificial intelligence (AI) and machine learning (ML), are an evolution above and beyond what traditional human-centric advising has been. That is somewhat true, though the term "Robo-advisor" conjures up the visual of a human reveling in financial advice. According to Jung et al. (2019), Robo-advisors are defined as tech-based assistants with intelligent, interactive user support. These tools allow financial advisors to virtually shepherd their customers through a financial advisory process using information technology. This is the story of Robo-advisors, a story set in the 21st century with unprecedented advances in computing power, data analytics, and algorithm development to process large datasets efficiently (Huang et al., 2022). They graduated from their algorithmic origins that offered rudimentary asset allocation assistance and have since evolved into incredibly advanced robots that utilize ML to provide intricate, personalized investment strategies in order to navigate these complex market conditions and offer specific financial advice to a widespread investor base. A common theme emerges if we look at the evolution of Robo-advisors: cost effectiveness. With the cost-conscious, fee-sensitive consumers of today in mind, Robo-advisory platforms provide financial advice to individuals for a fraction of what it would typically cost in fees if the advice had been sought via traditional human advisors (D'Acunto et al. 2019). There is something palpable when it comes to the democratization of financial advice, and it offers a way for certain demographics, who have historically been excluded, to be part of the journey towards creating wealth.

However, the benefits of Robo-advisors went beyond just cost-saving (Huang and Rust, 2017). Their power is anchored in a message of democratization and desegregation. Robo-advisors operate in real-time in a digital world, irrespective of geographical locations and time limitations, serving need-driven individual investors 24/7 as per the dynamics and complexity of the global interconnected trading system (Huang et al., 2022). This inclusivity appeals to a wide swath of demographics, presenting a case to not only experienced investors but also younger generations looking for financial advice in an internet-first world. The root of the Robo-advisory appeal thrives in the land of personalization and customization, which, if one were to take a deep dive into the real money-making mission, is part of the broader trend in open banking. Indeed, ML algorithms fueled by terabytes of data enable these digital platforms to decode intricate patterns in investor behavior, preferences, and financial aspirations (Jung et al., 2018). The outcome is a completely custom investment strategy with a vocabulary carefully tailored to the specific aims and degree of risk that exist for investors as individuals—the antithesis of financial advice as one-size-fits-all. Transparency is a key theme in this delicate dance between technology and finance. Trust is the name of the game in any financial advisory service, and Robo-advisors—praised for their transparency in how they compose investment strategies and build their fees—are among those that do it right. The transparent communication of algorithms and decision-making mechanics further helps to build trust between the consumers and their digital financial advisors (Roh et al., 2023). Yet, transformation comes with its trials. One of the challenges faced by Robo-advisors has been the perceived lack of human connection—it's a little intangible. Issues that need to be addressed so as to reveal the opportunities that point a way where technology meets human values.

The chapter will look into the factors influencing consumer purchase intentions and what is promoting them, and shed light on opportunities where technology can follow the basic tenets of most otherwise antiquarian financial services.

11.2 THE EVOLUTION OF ROBO-ADVISORS

1. **Inception and Early Pioneers:** The development of Robo-advisors originated in the aftermath of the 2008 financial crisis. This led to the idea of Robo-advisors. The need for more easily reachable and affordable investment solutions led to the proposal of automated advisory services.

2. **The Dawn of Algorithmic Asset Allocation:** Robo-advisors launched in their infancy with the automated asset allocation algorithm. Mathematical models were employed by these platforms to find the optimal investment portfolios for users, accounting for their risk tolerances, time horizons, and financial goals. The idea was to simplify the investment experience and get rid of all the clutter that came along with typical portfolio management.

3. **Machine Learning Integration:** The first iterations of behavioral finance largely focused on making micro-level observations and incorporating them into standard economic models. This involved doing things like asking people about their cognitive biases, reactions to financial events, so forth, then creating some final dichotomous measure thereof, linking this with savings rate, for example, and placing it into regression analysis. These results often troublingly turned mostly insignificant due to lax data collection practices. The advent of AI has seemed like a savior in that the data-driven approaches it allows promise greater comprehensiveness of independent, interpretable insights.

4. **Personalization and Customization:** However, when Robo-advisors really took a leap forward in their evolution, they began integrating personalization and customizability into the mix. ML algorithms were developed to analyze not just financial data but also to interpret individual investor preferences, actions, and life goals. This is an area where Robo-advisors could make investment strategies tailored with a finesse that resembled the nuanced advice of human advisors.
5. **Investor Type Services: Expanding Beyond Investing:** Recognizing consumers' complex financial demands, Robo-advisors expanded their offerings beyond managing investments. This gives a clue to the origins of Robo-advisors but also begs the question: why did they expand into goal-based financial planning, tax optimization, and retirement calculations? Why have they become an end-to-end platform for do-it-yourself personal financial planning?
6. **Hybrid Models in Collaboration:** While some services were entirely automated, with increasing recognition of the value of human expertise, others eventually offered a human advisory layer in addition at an affordable price. This was a bot-slash-advisor model, which means trying to combine the algorithmic efficiency with the human intuition of an advisor.
7. **Ethical and Responsible Investment:** Robo-advisors brought ESG investing into the mainstream. Ethical and responsible investing is an area that is becoming popular among investors who want to ensure their investment decisions reflect their moral outlook, and many platforms are all over this space.
8. **Regulatory Adaptation:** Of course, Robo-advisors did not explode out of the gates in a vacuum; there were additional societal and regulatory trends at play. Here are three key influencers. Once the platforms matured, regulatory bodies worldwide began to adapt their regimes to address this new frontier for potential risk and compliance challenges. On one side, regulators, and on the other, Robo-advisory platforms focus on compliance measures and improvements in investor protection.

The narrative of the Robo-advisor is one about evolution, creativity, and the rapid reflexes with which they react to the needs and wants of modern consumers today. So Robo-advisors have stepped in as the new sandpaper of investing, providing bespoke digital solutions in an unprecedented, ethical manner, and helping to resolve a degraded world financial ecosystem by empowering each investor with easy access to a portfolio that works for them as we transform our region into a healthier and more involved financial ecosystem. As we scan through these tiers, from Robo-advisors to intelligent advisors, the evolution is underway, and every page reveals more about how technology blends with financial intelligence.

11.2.1 FACTORS AFFECTING CONSUMER PURCHASE INTENTIONS

This demand is coming from a combination of influences, and understanding the factors that influence consumer decisions to use digital (Robo) advisory services requires an understanding of the depth and breadth of the modern-day financial advisory market. We then get into a comprehensive discussion outlining the various important parameters that drive consumers to embrace the services related to Robo-advisory.

1. **Cost Efficiency:** The most obvious consumer consideration is cost, which also gets lauded for its ability to provide financial advice significantly cheaper than traditional human advisors (Park et al., 2016; Faubion, 2016). This is an easy sell to a consumer base that is increasingly aware of how their investment outcomes erode as fees pile on. As for the bottom line, it will probably continue to be driven most directly by cost: Robo-advisory services offer unbeatable cost efficiencies, as ultimately that is the top priority for just about anyone, outside of maximizing returns per margin of capital and minimizing expenses.
2. **Accessibility and Inclusivity:** One of the key disruptions Robo-advisors have presented to the financial advisory industry is availability—24 hours a day, digitally. This level of accessibility caters to the accelerated pace of contemporary life and is also above geographical boundaries (Huang et al., 2022). People who have essentially been shut out by more traditional financial institutions are drawn to the inclusivity of Robo-advisory services, and the

model is attractive to a wide demographic, including younger investors.

3. **Personalized and Customized:** The continued allure of personalized financial guidance can act as a magnetic force, pushing consumers to turn to the likes of Robo-advisors. The integration of ML algorithms allows these platforms to analyze the preferences, risk tolerances, and financial aspirations of an individual and then create an investment strategy accordingly. The robots' ability to provide tailored advice for individual situations increases the attractiveness of the services provided, providing a level of personalization previously offered only by human financial advisors.

4. **Transparency and Trust:** Trust is paramount in the world of Robo-advisors, and it all starts with transparency. Clear communication of investment strategies, fees, and the underlying algorithms contributes to transparency in an era when consumers are acutely aware of the importance of knowing where their money is being invested. It is critical that trust be built because consumers will be leaving their financial well-being in the hands of an algorithm that must have a simple, comprehensible relationship with its users (Roh et al., 2023). Two different aspects are pricing, which means the fee for the service, and trustworthiness, which refers to the Robo-advisor's reliability and credibility (Salo, 2017).

5. **Confidence with and Ease Using Technology:** Consumer tech savviness and willingness to navigate digital services are at the core of Robo-advisor adoption. The shift to more digital natives may be a little easier, as they are already used to living a good chunk of their lives online; hence, using something like a Robo-advisory platform might be something that comes more intuitively to them. Yet, it is also important to ensure that these platforms are suitable for use among all users, due to the fact that previous activity indicators are mixed with different levels of user technological comfort, a point particularly relevant in combating digital inequality.

6. **Educational Initiatives:** Consumer awareness and understanding of Robo-advisory services are key to their purchase intention. Educational programs explaining how Robo-advisors work, what

algorithmic financial guidance can accomplish, and addressing common concerns assist the investor in making a well-informed decision. The government should take the first steps to hold campaigns, seminars, and awareness programs. Furthermore, according to Bhatia et al. (2020), regulators are planning that every financial advisory enterprise should have to share this information regarding Robo-advisory services with their current clients.

7. **Regulatory Environment:** The regulatory environment in which Robo-advisors operate plays a significant role in consumers' confidence as well as purchase intentions. Growing investor confidence in secure, reliable exchanges for grueling assets and investments is created by a network of regulators who emphasize customer security, data privacy, and ethical compliance. Users will feel better about using platforms that go beyond the minimum and actively conform to and announce their compliance with regulatory standards.
8. **Social Proof and Word of Mouth:** Word of mouth in the form of peer or industry recommendations and testimonials can dramatically influence consumer perceptions and purchase intent. Robo-advisory platforms gain credibility when their customers are not only using the platform but also recommending it. Many times, consumers look for reassurance in the experiences of others, and if you have good social standing, it's enough to sway anyone.
9. **Ethical and Social Impact:** As greater attention is paid to environmental, social, and governance (ESG) factors having an influence on stock performance, and more demand emerges for sustainable investing, consumers are also wanting to align Robo-advisory platforms with their values. For socially conscious investors, a platform that weaves ESG criteria into their investment approach could check a box that represents the ethical consideration of purchases.
10. **Combination Models with Human Touch:** Hybrid models that provide human advisory support also take care of the human touch needed by consumers in their financial interactions. To learn and understand what people want, enterprises must utilize platforms that facilitate a mix of algorithm-powered convenience and a

touch of helpful hand-holding for such a wide range of needs, thus acknowledging the central role relationships play in more impactful financial decisions.

Consequently, a plethora of economic, psychological, as well as technological, factors interact to shape consumer purchase intentions about Robo-advisors. The ability to navigate these factors successfully by Robo-advisory platforms is crucial for their broad adoption and establishing a symbiotic relationship with consumers in the new era of financial advisory services.

11.2.2 CHALLENGES AND OPPORTUNITIES IN THE ROBO-ADVISORY LANDSCAPE

Given that Robo-advisors are redefining the financial advisory landscape, it is important to take a closer look at some of the opportunities and challenges that will influence their direction. In this deep dive, we investigate the complexities surrounding this world of challenges as well as explore some of the exciting opportunities on the horizon for them in a constantly changing financial landscape.

11.2.2.1 CHALLENGES

1. **Lack of Human Touch:** As Robo-advisors are a type of financial advisor that provides financial advice or investment management online and are similar to their traditional counterparts, there are concerns about investors utilizing these services and platforms (Phoon and Koh, 2017). Many financial decisions involve emotional and complex considerations that an algorithm may find challenging to understand. Two of these challenges arose as dealing with the emotional stress aspects of financial planning, together with trust, can be difficult in a realm that has always belonged to human advisors (Aw et al., 2023).
2. **Communication and Education:** Communicating the value Robo-advisors have mostly struggled at communicating their value propositions and features to users, so a simple website with pricing information will already put them ahead of the competition. Educational programs are crucial to helping the average user grasp complex algorithms. Making the language of financial

advice easier to understand, with a focus on holistic resources, is fundamental to building consumer trust and adoption.

3. **Regulatory Compliance:** Regulatory rigor is subject to specifics that differ from jurisdiction to jurisdiction and are evolving. Compliance with new regulations, potential legal challenges, and possible changes in the data protection laws only add to the complexities of working within this sector. The high-level responsibility of Robo-advisors is to maintain regulatory compliance to ensure consumer trust (Huang et al., 2022).
4. **Algorithmic Risks and Biases:** Risks are introduced when relying on algorithms and, subsequently, their biases or decision-making errors. Robo-advisors are very much still in their infancy, and they need to constantly tweak algorithms to minimize exposures and ensure that investment strategies adapt as markets do the same. According to Huang et al. (2022), it is essential to address the algorithmic risks to the credibility and performance of these platforms.
5. **Cyber Security Concerns:** Similarly, Robo-advisors are digital platforms themselves and hence vulnerable to cyber-security threats. Certainly, one of the major challenges any bank faces is securing sensitive financial data and improvising cybersecurity, which is never a destination but a journey. Strong cybersecurity is a prerequisite to assure concerned users about their financial privacy.

11.2.2.2 OPPORTUNITIES

1. **Hybrid Models:** Robo-advisors stand to benefit from blending human advisory help with automated services by offering hybrid models. By recognizing that the human element remains crucial in financial interactions, COBA is targeting consumers who want a touch of personalization to accompany their slick use of algorithms. These are hybrid models between technology and the personal touch of traditional financial advice services.
2. **Frequent Technological Developments:** As new technology continues to leapfrog, Robo-advisors can improve themselves

more and more. Advancements in artificial intelligence (AI) and ML, as well as data analytics, allow these platforms to offer more complex forms of personalized financial advice. This will allow Robo-advisors to continue serving the continuously changing needs of financial consumers (Jung et al., 2018).

3. **Expanding Beyond Traditional Investment Management Services:** Positioning themselves as comprehensive financial planning tools helps expand the scope of consumer financial needs they are able to serve. As Hildebrand and Bergner (2021) argue, Robo-advisors are set to improve financial advice quality, which is expected to be both more reliable and transparent for consumers.

4. **Ethical Investing or Responsible Investing:** Such a shift only favors Robo-advisors who adopt responsible and ethical investing, as those with an eye on impact are best targeted towards genuine socially conscious consumers. Setting investment strategies that take into account ESG criteria can appeal to people who want to invest their money as sustainably and ethically as possible. According to Belanche et al. (2019), Robo-advisors are anticipated to help customers make better investment decisions and lower the costs of advice.

5. **Educational Initiatives:** Education outreach campaigns help demystify Robo-advisory services and enable financial literacy, fueling a rise in adoption. Robo-advisors can solve the user education challenge and create a more informed user base by investing in educational resources and clear communication.

6. **Cost Cutting:** The cost disruption that Robo-advisors propose indicates why such financial and banking firms view Robo-advisors as symbiotic with their enterprise operational strategies (Isaian and Oggero, 2022).

In summary, the opportunities created by technology improvements, evolving consumer demands, and business model innovations counteract some of the challenges faced by Robo-advisors. This is a landscape that Robo-advisors must navigate with a solution that, unlike the market, evolves and adapts dynamically while also understanding the subtleties of the rapidly evolving financial ecosystem.

11.3 CONCLUSION

With this, we come to the end of our venture into the world of Robo-advisors and their implication on consumer purchase intent—a confluence of technological innovation, financial evolution, and evolving consumer behaviors. This trajectory through the evolution of Robo-advisors highlights a narrative that extends beyond that of plain and simple disruption; it is a changing, transformational process by which people get financial advice. As Robo-advisors have developed from an original idea to the advanced platforms we now see, AI and data analytics technology have evolved by leaps and bounds (Sironi, 2016). As discussed, digital financial advisors have not only automated investment activities but have also opened up a new era of democratization by offering accessible and cheaper financial advice customized for single customers. What influences consumer purchase intentions is a patchwork of perspectives. The five basic pillars for the adoption of Robo-advisory services are cost-effectiveness, accessibility, personalization, transparency, and trust. Hybrid models, leveraging efficiencies by providing a mix of services, ethical investing, a worker-friendly approach to finance, and leveraging collaborative ventures to close gaps met the challenges faced by these platforms—from the heartless touch problem to growing regulatory complexities. In the future, the horizons of Robo-advisors are inviting them to intersect technology with finance like never before. Taking cues from these signals, the infusion of cutting-edge technology, scalability beyond investing, and a shift towards high-touch individualization on a mass scale point to highly personalized paths devoid of inefficiency. Whether it is collaborating with traditional financial institutions, prioritizing responsible investing, sustainability, or aiming for financial inclusion, all these are signs of a greater societal shift towards a more inclusive and ethical financial ecosystem. With the nuanced interplay of algorithms and human touch, the importance of continual learning and dialogue, and a shifting regulatory landscape, mines create a developing process with varied depth in deep waters. Robo-advisors who are able to be nimble, transparent, and focused on giving value to the broad spectrum of clients they serve will win in this challenging new environment. However, modern marketing and operational practices in the banking and finance industry are almost inseparable from digital technologies that integrate AI and customer data (Quach et al., 2022). As the body of knowledge from fields such as behavioral economics expands,

Robo-advisory services can assist decision-making and improve investment behaviors. Robo-advisors will thus, in the long term, help investors reduce behavioral biases compared to standard traditional advisory solutions offered by asset managers (Uhl and Rohner, 2018).

As we finish up with Robo-advisors and consumer purchase intentions, it is evident that the story is nowhere near finished. The following pages are an invitation to forge a new future where imagination and cooperation take us beyond the horizon of possibility in financial planning. Financial services businesses (and the regulators overseeing them) that navigate this terrain with foresight, transparency, and a useful sense of context for their customers will shape the next generation of precepts in financial services (Silva, 2019). While Robo-advisors comprise disruptors in this always-on story, they are more so the creators of a kinder, gentler, and techier financial world on the horizon.

11.3.1 LOOKING AHEAD TO ROBO-ADVISORS

This graphic shows the future landscape of robots as a result of technological advancements, evolving consumer preferences, and tactical innovation. The following is an exploration into some of the areas enabling opportunities and trends that might characterize the next phase in the evolution of Robo-advisory services.

1. **Using Modern Technologies:** Robo-Advisors and Modern Technologies Build on the Future of a Generation. As AI, ML, and predictive analytics continue to advance, Robo-advisors will soon deliver even more personalized and responsive financial advice than today. Natural Language Processing (NLP): The addition of natural language processing can also improve communications, leading to a more user-friendly and familiar interaction with Robo-advisors.

2. **Expansion Beyond Investments:** As we look even further ahead, the terrain widens to include newer and broader Robo-advisor horizons. We are already seeing these change into full financial wellness tools and will likely continue to morph into a suite of services, from debt management all the way to budgeting support and holistic financial planning, strategically making Robo-advisors the primary access point for anything financial for a client, from account checking to private equity.

3. **Hybrid Models and Human-Enablement:** The hybrid model, which combines the efficiency of algorithms with human advisory support, is likely to become more prevalent. Because people are pretty adept at sensing when information is relevant to them personally, Robo-advisors could work together with human advisors in a very streamlined way that lets you find the right balance of automated and face-to-face service on your terms. This hybrid model aims to cater to a wide variety of consumer tastes, blending the technological with the human assets of financial experts.
4. **ESG Ethical and Sustainable Investing:** With the use of a Robo-advisor, developers are developing patterns using collected trends and massive data. In turn, this will allow experts in this area to design superior algorithms to improve the predictions of reasonable investment decisions. According to Bhatia et al. (2020), using big data will significantly improve the risk-return trade-off calculation for the portfolio optimization offered by advisory services.
5. **Customization at Scale:** It will likely be future generations of Robo-Advisors that are able to offer such a level of customization through the use of big data and advanced analytics. The ability to analyze big data in real-time supports rapid changes in the market and investor interests. Individualized investment strategies that are designed to fit the user's or intended user's specific financial goals and risk threshold have now become not only possible but even integral to the Robo-advisory journey.
6. **Enhanced User Experience:** The future of the user experience is likely to be more streamlined in the space of upcoming Robo-advisors. These will cohesively blend intuitive interfaces, interactive tools, and educational resources to inform and upskill end users on financial literacy and decision-making. The goal is to strengthen the user journey and ensure that it is not only more transparent but also empowering—an ideal scenario conducive to increased public trust in Robo-advisory services.
7. **Regulatory Frameworks and Standardization:** As the space matures, one would expect that regulatory frameworks will take shape and become more standardized in different jurisdictions. Smart regulations featuring updated investor protection,

data privacy, and law enforcement standards will set a lasting groundwork for growth inside the sector. This also plays a role in greeting the consumer with more familiarity, an arguable advantage leading to increased comfort and favoring the large-scale use of pesticides.

8. **Working with Credit Institutions:** Collaborations between Robo-advisors and incumbent financial players are expected to grow. They can enable high-end Robo-advisors to access sophisticated technology and legacy institutions to brand both ends, developing hybrid financial ecosystems that will appeal to the various demographics of customers. These partnerships could be in the form of joint ventures, strategic alliances, or acquisitions, leading to a restructuring of the competitive landscape.
9. **Focus on Financial Inclusion:** Financial inclusion is likely to become a priority among Robo-advisors in the coming future landscape. Robo-advisors have the potential to help those who have been overlooked gain access to financial advisory and investment opportunities by using digital platforms and new methods. The democratization of financial services is also broadly aligned with the societal interest in inclusivity and accessibility.
10. **Evolution of Cyber Security Continues:** As today has seen the largest ever reliance on digital platforms, so will we need to adapt to the increased security vulnerabilities that come with it. As the industry matures, the future of Robo-advisors will likely place greater emphasis on instituting a security-first policy to protect financial data and secure against more sophisticated cyber threats. As we handle more and more sensitive financial consumer data, proactive cybersecurity measures will be key to ensuring end-consumer faith in the safety of their data.

In summary, the Robo-advisor world of tomorrow is likely to be a more exciting, disruptive place, one shaped by the convergence of emerging technologies and new services and partnerships. As these digital financial advisors experience continued evolution, the potential of their reach within financial services and their ability to change the behavior of wealth management may be more than simply very deep. If they play their cards right, it could presage a future where financial planning is personal, ubiquitous, and fully embedded in our lives.

KEYWORDS

- **consumer purchase intentions**
- **financial inclusion**
- **personalization**
- **robo-advisors**
- **trust**

REFERENCES

Abraham, F., Schmukler, S. L., & Tessada, J. (2019). Robo-advisors: Investing through machines. *World Bank Research and Policy Briefs* (134881).

Aw, E. C. X., Leong, L. Y., Hew, J. J., Rana, N. P., Tan, T. M., & Jee, T. W. (2023). Counteracting dark sides of Robo-advisors: Justice, privacy and intrusion considerations. *International Journal of Bank Marketing*.

Belanche, D., Casaló, L. V., & Flavián, C. (2019). Artificial intelligence in FinTech: Understanding Robo-advisors adoption among customers. *Industrial Management & Data Systems, 119*(7), 1411–1430.

Bhatia, A., Chandani, A., & Chhateja, J. (2020). Robo advisory and its potential in addressing the behavioral biases of investors—A qualitative study in Indian context. *Journal of Behavioral and Experimental Finance, 25*, 100281.

D'Acunto, F., Prabhala, N., & Rossi, A. G. (2019). The promises and pitfalls of Robo-advising. *The Review of Financial Studies, 32*(5), 1983–2020.

Faubion, B. (2016). Effect of automated advising platforms on the financial advising market. (Bachelor's thesis, Sam M. Walton College of Business, University of Arkansas, Fayetteville, AR). Available at: https://core.ac.uk/download/pdf/72841372.pdf (accessed on 11th August 2025).

Hildebrand, C., & Bergner, A. (2021). Conversational Robo advisors as surrogates of trust: Onboarding experience, firm perception, and consumer financial decision making. *Journal of the Academy of Marketing Science, 49*, 659–676.

Huang, M. H., & Rust, R. T. (2017). Technology-driven service strategy. *Journal of the Academy of Marketing Science, 45*(6), 906–924.

Huang, R. H., Wang, C. C., & Zhang, O. X. (2022). The development and regulation of Robo-advisors in Hong Kong: Empirical and comparative perspectives. *Journal of Corporate Law Studies, 22*(1), 229–263.

Isaia, E., & Oggero, N. (2022). The potential use of Robo-advisors among the young generation: Evidence from Italy. *Finance Research Letters, 48*, 103046.

Jung, D., Dorner, V., Weinhardt, C., & Pusmaz, H. (2018). Designing a Robo-advisor for risk-averse, low-budget consumers. *Electronic Markets, 28*, 367–380.

Jung, D., Glaser, F., & Köpplin, W. (2019). Robo-advisory: Opportunities and risks for the future of financial advisory. In *Advances in Consulting Research: Recent Findings and Practical Cases* (pp. 405–427).

Park, J. Y., Ryu, J. P., & Shin, H. J. (2016). Robo advisors for portfolio management. *Advanced Science and Technology Letters, 141*(1), 104–108.

Phoon, K., & Koh, F. (2017). Robo-advisors and wealth management. *The Journal of Alternative Investments, 20*(3), 79–94.

Quach, S., Thaichon, P., Martin, K. D., Weaven, S., & Palmatier, R. W. (2022). Digital technologies: Tensions in privacy and data. *Journal of the Academy of Marketing Science, 50*(6), 1299–1323.

Roh, T., Park, B. I., & Xiao, S. S. (2023). Adoption of AI-enabled Robo-advisors in FinTech: Simultaneous employment of UTAUT and the Theory of Reasoned Action. *Journal of Electronic Commerce Research, 24*(1), 29–47.

Salo, A. (2017). Robo advisor, your reliable partner? Building a trustworthy digital investment management service (Master's thesis).

Silva, P. M. (2019). Robo-advising: Unfolding the risks. In *3rd International Conference on Computational Finance (ICCF2019)* (p. 214).

Singh, I., & Kaur, N. (2017). Wealth management through Robo advisory. *International Journal of Research–Granthaalayah, 5*(6), 33–43.

Singh, J., Arnold, T., Brady, M., & Brown, T. (2019). Synergies at the intersection of retailing and organizational frontlines research. *Journal of Retailing, 95*(2), 90.

Sironi, P. (2016). *FinTech Innovation: From Robo-Advisors to Goal Based Investing and Gamification*. John Wiley & Sons.

Song, S. Y., & Kim, Y. K. (2022). Factors influencing consumers' intention to adopt fashion robot advisors: Psychological network analysis. *Clothing and Textiles Research Journal, 40*(1), 3–18.

Uhl, M. W., & Rohner, P. (2018). Robo-advisors versus traditional investment advisors: An unequal game. *The Journal of Wealth Management, 21*(1), 44–50.

CHAPTER 12

Analyzing the Global Research Landscape of AI in Electronic Human Resources Management: A Bibliometric Approach

ARCHANA SINGH,[1] GIRISH LAKHERA,[1,*] and MOHIT KUKRETI[2]

[1]*Department of Management Studies, Graphic Era (Deemed to be University), Dehradun, Uttarakhand, India*

[2]*Department of Business Studies, University of Technology and Applied Sciences (UTAS), College of Applied Sciences, Ibri, Oman*

*Corresponding author

ABSTRACT

This study undertakes a bibliometric exploration of the global research landscape at the nexus of Artificial Intelligence (AI) and Electronic Human Resource Management (E-HRM), spanning publications from 1988 to 2023. Drawing upon 274 journal articles indexed in Scopus and Web of Science, the investigation delineates the intellectual architecture, thematic trajectories, and geographic dispersion of scholarship in this domain. Findings reveal an exponential growth of publications post-2017, with the United States, China, and India emerging as leading contributors, and Aston University serving as a prominent institutional hub. Keyword co-occurrence and co-citation analyses identify five dominant clusters—AI integration, HRM innovation, machine learning, reinforcement learning, and recruitment—underscoring the interdisciplinary evolution of the field. Despite this expansion, critical gaps persist, particularly

Robo-Advisors and Artificial Intelligence in Human Resources Management: Revolutionizing HR.
Sanjay Taneja, Swati Gupta, Girish Lakhera, Mohit Kukreti, and Ercan Özen (Eds.)

DOI: 10.1201/9781779641410-12

regarding sustainability, ethics, and qualitative validations of E-HRM frameworks. The study advances both theoretical and practical insights by highlighting the transformative role of AI-driven HR analytics, talent acquisition systems, and e-learning platforms in shaping organizational efficiency. For scholars, the research maps fertile avenues for future inquiry, while for practitioners, it offers strategic directions to harness AI-enabled E-HRM for agility, inclusivity, and competitive advantage in dynamic labor markets.

12.1 INTRODUCTION

The integration of artificial intelligence (AI), identified by the Society of Human Resource Management (SHRM) as a leading technological trend and a key catalyst for the future of HRM, underpins the progress in decision-making capabilities (Aman et al. 2022). As a result, although being touted as a potent tool for HRM, academic study on the topic is not very comprehensive (Ahsan et al. 2021). The following research topics for this study can be established thanks to the relationship between AI and EHRM. The first research issue relates to earlier AI considerations and difficulties. This study intends to map the literature through a bibliometric analysis that gives a reflection on the progression of AI into EHRM from 2010 to 2020 in light of the increased attention and importance AI and EHRM have been achieving. The study analyzes the intellectual structure of the subject and identifies key research gaps while providing a complete picture of the research that has been done over the past three decades on the topic of "Human Resource Management (HRM) and Artificial Intelligence" in the Electronic Human Resource Management domain. The research identified significant contributors to the study are Lecture Notes in AI and Lecture Notes in Bioinformatics), geographical areas under which research in the discipline is focused (the leading country: USA), and the universities emphasizing AI and HRM (leading university: Aston University). The knowledge base on EHRM and AI subjects may be categorized into five main themes: HRM and AI, Recruitment, Talent Acquisition, Training, and E-Learning. These themes were also found in the technical advancement of HRM. The need for more research was then made clear by highlighting fundamental research gaps. Leaders in the industry and university scholars would be able to get new, advanced insights into this developing subject thanks to this study. Such knowledge would be useful in creating strategies for HRM. The study will assist in the creation of policies to support the growth of electrical human resources (HR) management.

12.2 MATERIAL AND METHOD

By combining HRM and IT procedures, e-HRM seeks to improve employee performance by streamlining people management inside businesses (Bondarouk et al., 2011; Ruel et al., 2014). The study's data was taken from the Scopus database up until October 2023, after which abstracts from every journal pertaining to HR were examined and sorted. There are a few restrictions to take into account, though. Because book chapters and conference papers are not as frequently included in literature-based data, relying only on the Scopus and Web of Science databases may have resulted in the exclusion of pertinent publications.

Even with strict adherence to policies and procedures, there may always be some subjectivity involved in the classification of publications. To ensure the inclusion of all pertinent publications, it is important to note that no journals were excluded based on quality indicators like rankings or the H index. Since journal articles are regarded as trustworthy sources of verified knowledge, this study only looked at journal articles (Podsakoff et al., 2005). Book chapters and conference papers were not included because of their restricted accessibility and laxer review procedures (Jones et al., 2012). 45 results were found from a search conducted from 1995 to 2023 using the keywords "E-HRM" and "AI" in the Scopus and Web of Science databases.

12.2.1 CITATION INDEX SELECTION

Using information obtained from Clarivate Analytics Web of Science and Scopus, this review expanded upon the work done by Bartolinni et al. and Zhanng et al. The incorporation of multiple databases was considered necessary to guarantee thorough coverage of the subject, thereby reducing the coverage and area limitations that arise from depending solely on one database. The integration of data from both sources broadened the study's focus and improved the validity of its conclusions. With their accurate description of these two databases as "world-leading and competing," Zhu and Liu draw attention to their importance in scholarly research. In contrast,

Scopus serves as a comprehensive bibliographic database covering a broad spectrum of literature in fields such as science, technology, medicine, and social sciences.

12.2.2 SELECTION OF KEYWORDS

Searching databases for relevant publications using keywords is one way to find relevant literature. A good way to deal with this problem is to carry out a thorough keyword search inside academic journal articles about the particular subject. While determining the final keywords, researchers took into account a number of synonyms for the main term "E-HRM." They expanded the search to include related terms with different endings by using the asterisk (*) wildcard in their queries.

The same keywords were used for searches in both databases to guarantee consistency. Data was obtained by searching Scopus with the following term: KEY-ABS-TITLE ("E-HRM" "electronic human resource management") and LIMIT-TO (PUBSTAGE, "Information technology").

12.2.3 DATA COLLECTION AND PROCESSING

While there were no restrictions on the range of years considered due to the limited accessibility of E-HRM articles, the search criteria focused on including articles from India, review articles, and those written in English. It's noteworthy that there was no temporal constraint imposed, as the review aimed for a comprehensive exploration and analysis of trends spanning from the inception of E-HRM research to the present. However, for accuracy, only the titles of the documents were examined. As a result, a total of 274 publications were obtained after removing duplicate entries using the R package.

12.2.4 BIBLIOMETRIC ANALYSIS AND SOFTWARE PACKAGE

The study made use of the bibliometrics R-package, an open-source application created by Aria and Cuccurullo that contains a number of tools for using the R programming language to perform quantitative bibliometrics research. Notably, an online interface application called Biblioshiny was added to the program with updates (version 2.0 and above). With this addition, bibliometric research output can be generated by non-programmers via an easy-to-use interface.

The investigation used VOSViewer and the R program to apply customized bibliographic criteria for WOS and Scopus datasets in order to perform bibliometric analysis.

12.3 FINDINGS

12.3.1 SYNTHESIS DATA

This research delved into the realm of Electronic Human Resource Management (E-HRM) and its influence on AI by examining literature published on the subject over the past 30 years, starting from 1988. The study looked at the historical evolution and changes in E-HRM research with the goal of comprehending its theoretical, conceptual, and social underpinnings. Figure 12.1 presents the extensive findings from the bibliometric analysis and gives a thorough synopsis of the research on E-HRM and its influence on AI. In order to perform the cluster analysis, a thorough review of the literature was carried out, and publications within the identified clusters were carefully examined to reveal relationships between them.

FIGURE 12.1 A data synthesis.

Source: Biblioshiny.

12.3.2 PERFORMANCE ANALYSIS

AI and HRM manifested through 274 manuscripts published in journals. At least 80% of these works were cited. The first document was released in 1988, and the theory has been in operation for the past 35 years. Over 30 years, AI and Human Research collected 1937 citations, with an average of 5 citations per document, indicating an average level of intellectual output. Annual trends in AI and HRM are shown in Figure 12.2 of the EHRM domain. The study of technology and AI has gained popularity since 2002, as the world has seen the effects of technological advancement. However, since 2008, there has been a consistent rise in Human Resource Management (HRM). Looking at early 2020 publishing patterns, this rise is predicted to continue. According to the data's suggested development trajectory, publications on AI and HRM began to take up in 2005 and showed a continual increase after beginning in 1988. After a brief decline, publications resumed their upward trend until 2016. In 2017, the growth rate markedly accelerated. After then, the number of articles published on the subject increased exponentially, with 66 papers published in 2020, 97 papers published in 2021, and 117 papers published in 2022. AI and HRM have shown exponential growth from 2017 until the present. The publishing figures for 2021–2022 indicate that the field is still expected to be as productive as it was the year before, if not more so. This expansion may be ascribed to the UN's support of SDGs, focus on enterprises to attract HRM, and subsequent increased attention on HRM.

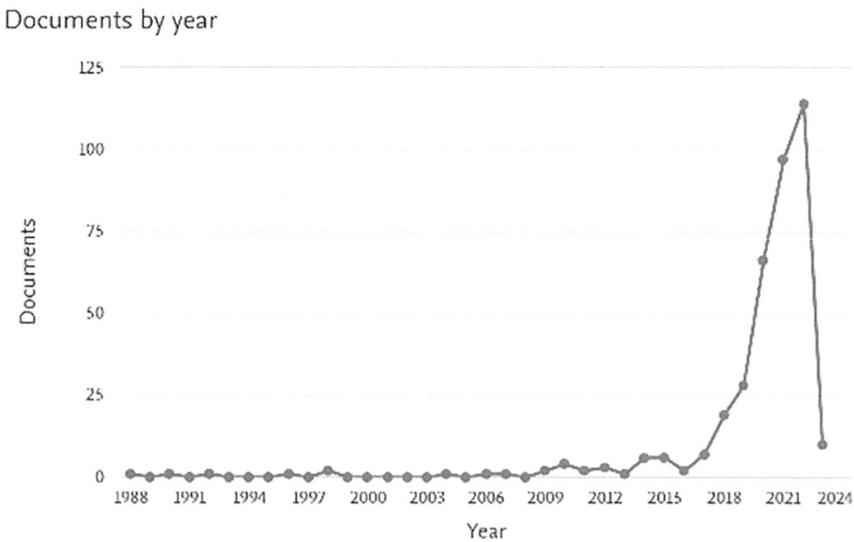

FIGURE 12.2 Distribution by year.

Source: Generated by Scopus.

12.3.3 THE MOST PROMINENT COUNTRIES

The U.S. is the leading contributor in publishing arrangements by country of origin, followed by China in second place and India in third, as shown in Figure 12.3. The U.K. and Germany, two European countries, are in 4th and 5th rank. Following this, Australia and France took sixth and seventh positions, respectively. Malaysia came in at number eight, followed by the United Arab Emirates at number nine, and Bahrain at number 10.

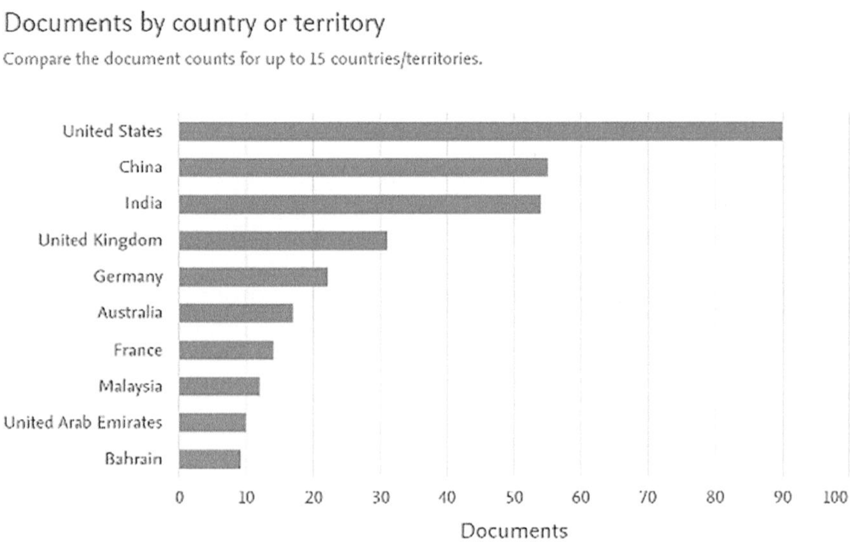

FIGURE 12.3 Most prominent country.

Source: Generated by Scopus.

12.3.4 THE MOST PROMINENT AFFILIATION

According to Figure 12.4, the segmentation of publications with affiliation indicates that the top ranks were led by Aston University in HRM and AI research. The University of New Castle Australia,

Carnegie Mellon University, Amity University, Aston Business School, Applied Science University, Pennsylvania State University, University of Houston, West Virginia University, and Clemson University, are the best contributors in the area.

12.3.5 THE MOST PROMINENT AUTHORS

The top author is Bhudwar. P Second, Malik A. has seven Scopus-indexed publications. Next, Cagan J. has five Scopus-indexed papers. 4th in a row is Gyory, J.T. with five Scopus papers In the Scopus database, Kotovsky, K., is in fourth place with five publications, followed by MC Comb., C., who has four publications. The majority of other well-known authors in the area depended on partnerships (CI > 1) to produce co-authored work as showed in Figure 12.5.

12.4 INTELLECTUAL STRUCTURE

12.4.1 KEYWORDS NETWORK ANALYSIS

Based on the network map of word co-occurrences created, we examined the terminologies relevant to the role of AI in EHRM in this section. In 377 investigations on AI function in EHRM, 1115 keywords were discovered, with numerous occurrences of 5 and a threshold of 24. A co-occurrence network was created using VOSviewer, as shown in Figure 12.6.

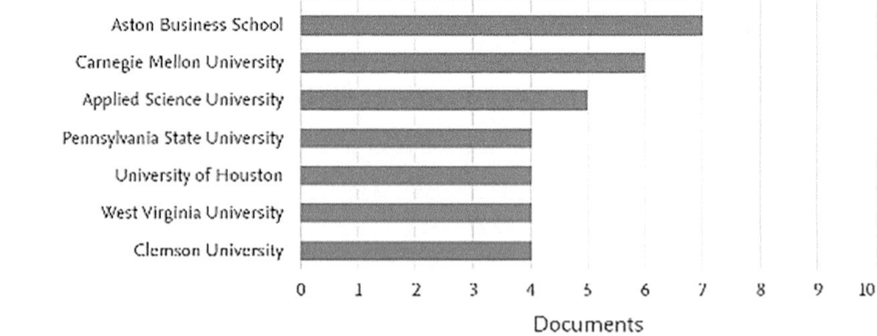

FIGURE 12.4 Distribution by affiliations.

Source: Generated by Scopus.

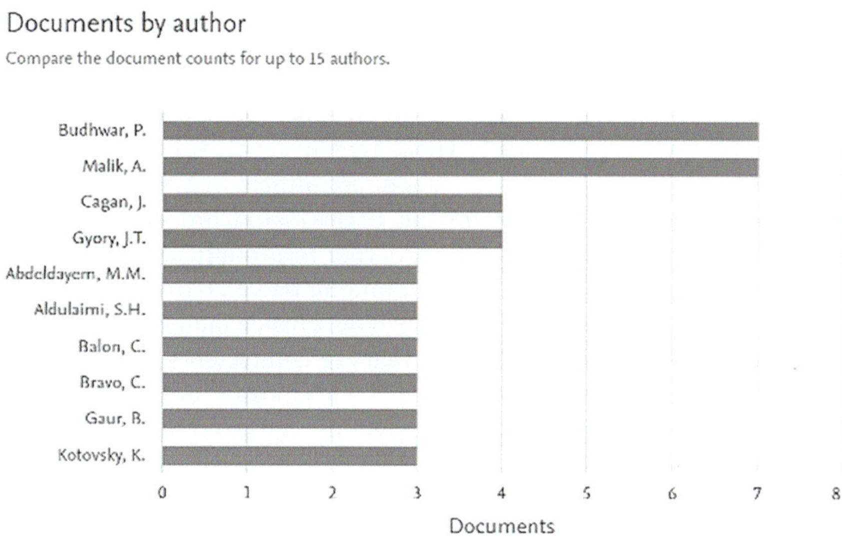

FIGURE 12.5 Most prominent authors.

Source: Compiled by Authors.

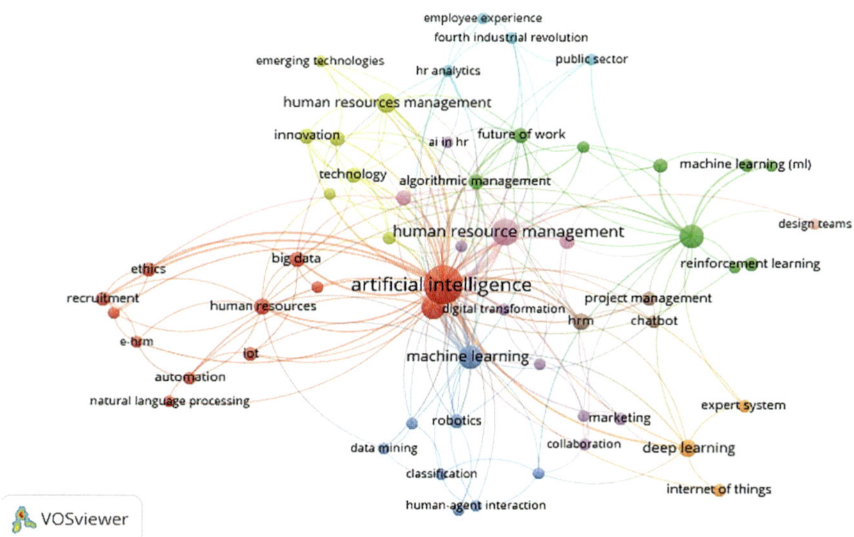

FIGURE 12.6 Keywords co-occurrence network.

Source: Generated by VOSviewer.

Five clusters constructed using VOSviewer and based on Fig. 6 have been produced. When viewing the map, nodes of the same color form clusters, and the intensity of the association between nodes in a cluster is shown by the shorter distance between them. Artificial intelligence (AI) is represented by Cluster 1; HRM by Cluster 2; ML by Cluster 3; reinforcement learning by Cluster 4; and recruitment by Cluster 5. Based on the analysis of the AI role in the EHRM keyword co-occurrence network, Figure 12.6 shows the top AI role in a publication related to the EHRM in terms of frequency and total link power. Co-word cluster 1 – Artificial Intelligence (Red network): This cluster includes topics like ML, big data, hiring, IoT, and robots that are connected to AI in human resource management. Co-word cluster 2: This cluster includes elements of HRM (purple network), such as innovation, technology, industry 4.0, algorithmic management, and ethics. The third co-word cluster has ML as a connection (green network). AI, deep learning (DL), HRM, and the future of networks are all included in this. Reinforcement learning is the fourth co-word cluster (blue network). This cluster focuses on the DL and algorithmic management aspects of HRM, which were assigned to the second cluster. Cluster 5: Recruitment.

12.4.2 THE CO-CITATION ANALYSIS

The use of co-citation to identify the affiliation of authors, academic works, and works in each field of study has proven beneficial (Osareh, 2009). Based on a criterion of 6 and a minimum of 2 citations, 15,180 citations, and a cited reference, the reference co-citation analysis on 274 documents was conducted. Five mentioned references need to be chosen. Figure 12.7 displays the outcome. Research clusters appear when a pair of articles is cited by several writers. The works that are referenced together in these clusters typically cover the same ground. Three clusters were developed as a result of the co-citation study, as shown in Figure 12.7. Red, green, and blue accents are used to draw attention to the cluster.

12.4.3 JOURNALS, AUTHORS, AND ABSTRACT KEYWORDS PLOT IN THREE FIELD

The interdependence of these elements is visually represented by the three-field plot, which connects author, publication title, and keyword

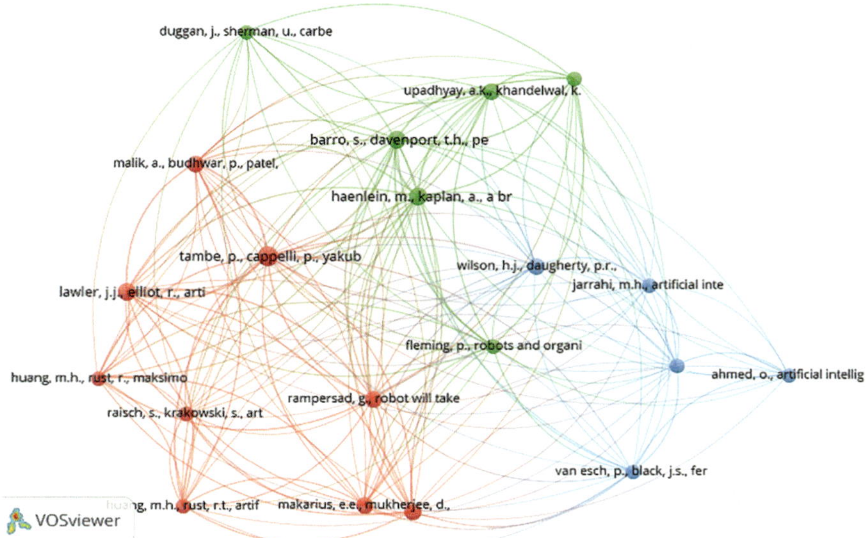

FIGURE 12.7 Co-citation analysis.

Source: Generated by VOSviewer.

taken from articles' abstracts with the help of a grey line. These lines draw attention to the connections between them by illustrating the relationship between authors, important abstract keywords, and the associated journals of publication. These connections are easier to understand quickly thanks to this visual representation. Furthermore, the size of every rectangle in each category (titles of publications, authors, and keywords) reflects the quantity of papers linked to that particular feature. Each bar's length serves as a visual cue about a topic's prominence in the literature at the moment, indicating the focus of the research. Users are able to evaluate the influence and area of emphasis of individual journals in the E-HRM domain by looking at the relationships among publication titles, authors, and journals. Researchers hoping to publish in prestigious journals will find this information useful. The visualization can assist in identifying areas in the existing literature that are less well-known or have gaps for those who are planning new studies. These observations serve as a guide when choosing study subjects and possible partners in understudied fields (Figure 12.8).

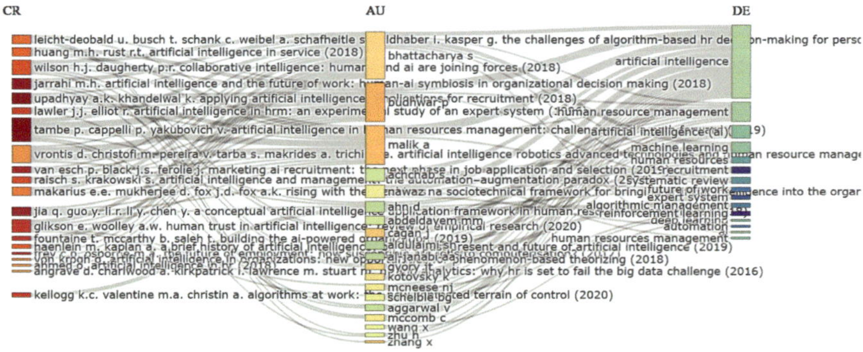

FIGURE 12.8 Three field plot.

Source: Biblioshiny.

12.5 DISCUSSION

Performance studies indicate that while this industry's publication count has grown consistently since 2010, a bigger growth occurred in 2022, which was followed by an exponential rise after 2018. In terms of AI's role in EHRM, the United States far outpaces China. The recent publication of numerous highly referenced articles suggests an increase in the number of academic publications on the function of AI in EHRM, as well as a change in the focus of some of this research. As a consequence, the industry seems to be developing significantly and moving forward. Unexpectedly, the US has the most authors and colleges, with China coming in second. The fact that the research suggests that the focus is still on how AI functions in EHRM, how they respond to various HR Strategies programs inside human resource elements, or whether particular personality traits influence narrowly defined behaviors is a potentially worrisome aspect. Less focus has been paid to HR changes and the importance of sustainability in AI and HR, which emphasizes how challenging it is to incorporate sustainability into HRM generally.

12.6 LIMITATION AND AVENUE FOR FUTURE STUDIES

Like any research project, this study has limitations even though it was carried out in great detail. First off, even with such a wide range of keywords, it's possible that some relevant details were unintentionally missed.

Furthermore, even though the Scopus and Web of Science databases offer extensive coverage, it's possible that they don't contain all pertinent research sources. As a result, research not included in these databases but still potentially providing important information might have gone unnoticed. Subsequent research endeavors may augment the scope of the study by broadening the range of keywords and integrating supplementary databases.

12.7 IMPLICATION

The themes that have emerged in relation to AI's role in E-HRM have been emphasized by this study. It verified the present understanding of E-HRM through a thorough Systematic Literature Review (SLR) and pinpointed particular areas requiring additional study for a more thorough grasp of the field. The research highlights the necessity of conducting more thorough investigation into the factors, actions, and competencies related to the function of E-HRM in AI. These results hold significance for researchers looking to broaden their comprehension as well as for professionals trying to improve their familiarity with the E-HRM idea.

12.8 CONCLUSION

The aim of this research was to analyze all of the studies that were available from the Scopus databases in order to fully investigate the role of E-HRM in AI. A methodical and comprehensive examination of the E-HRM literature was ensured by this methodological approach. This SLR focused on E-HRM, an important but dispersed type of information technology, and highlighted developments and trends in the field. Values, literature reviews, behavior, assessments, performance, features, and antecedents of E-HRM are just a few of the major themes that the study found in the literature on E-HRM. Notable research revealed that information technologies, employee performance, AI, and E-HRM practices are the four key competencies required for the role of E-HRM in AI. The literature's succinct explanations point to the necessity for more study to completely comprehend these elements. Furthermore, more research is required to reach a consensus on the causes of E-HRM in order to guide future procedures. The evaluation pointed out shortcomings in the study's methodology and empirical parts, which emphasized the need for more qualitative research. The significance of creating experimentally validated frameworks for the phenomenon of E-HRM was also underlined by the study. This chapter concludes with a thorough analysis of E-HRM in the context of AI, outlining the field's current state, its scope, and the areas in need of more investigation.

KEYWORDS

- **artificial intelligence**
- **bibliometric analysis**
- **electronic human resource management**
- **HR analytics**
- **machine learning**

REFERENCES

Ahsan, M. R. (2021). Challenges and opportunities concerning environmental goods (EGs): A case study and business perspective. *Society and Business Review, 16*(2), 184–200.

Al-Kasasbeh, A. M., Halim, M. A. S. A., & Omar, K. (2016). E-HRM, workforce agility and organizational performance: A review paper toward theoretical framework. *International Journal of Applied Business and Economic Research, 14*(15), 10671–10685.

Aman, A., Rafiq, M., Dastane, O., & Sabir, A. A. (2022). Green corridor: A critical perspective and development of research agenda. *Frontiers in Environmental Science, 10*, 982473.

Barboza, M. N. L., & Filho, E. J. M. A. (2019). Green consumption values in mobile apps. *Journal of International Consumer Marketing, 31*(1), 66–83.

Bissola, R., & Imperatori, B. (2013). Facing e-HRM: The consequences on employee attitude towards the organization and the HR department in Italian SMEs. *European Journal of International Management, 7*(4), 450–465.

Bondarouk, T. (2011). Theoretical approaches to e-HRM implementations. In *Electronic HRM in Theory and Practice* (pp. 1–20). Emerald Group Publishing Limited.

Bondarouk, T., Harms, R., & Lepak, D. (2017). Does e-HRM lead to better HRM service? *The International Journal of Human Resource Management, 28*(9), 1332–1362.

Bondarouk, T., Schilling, D., & Ruël, H. (2016). E-HRM adoption in emerging economies: The case of subsidiaries of multinational corporations in Indonesia. *Canadian Journal of Administrative Sciences, 33*(2), 124–137. https://doi.org/10.1002/cjas.1376 (accessed on 11th August 2025).

Boyack, K. W., & Klavans, R. (2010). Co-citation analysis, bibliographic coupling, and direct citation: Which citation approach represents the research front most accurately? *Journal of the American Society for Information Science and Technology, 61*(12), 2389–2404.

Burbach, R., & Royle, T. (2013). Levels of e-HRM adoption in subsidiaries of a US multinational corporation: The mediating role of power, politics, and institutions. *European Journal of International Management, 7*(4), 432–449.

Jones, J. T., Whitaker, M., Seet, P. S., & Parkin, J. (2012). Talent management in practice in Australia: Individualistic or strategic? An exploratory study. *Asia Pacific Journal of Human Resources, 50*(4), 399–420.

Podsakoff, P. M., MacKenzie, S. B., Bachrach, D. G., & Podsakoff, N. P. (2005). The influence of management journals in the 1980s and 1990s. *Strategic Management Journal, 26*(5), 473–488.

Ruël, H., & Bondarouk, T. (2014). E-HRM research and practice: Facing the challenges ahead. In *Handbook of Strategic E-Business Management* (pp. 633–653).

CHAPTER 13

Robo-Advisors in Investment Management

SAKSHI RATURI,[1] AMAR SINGH,[1] TANUJA GOUR,[2] and SHWETANK AVIKAL[3]

[1]*School of Commerce, Graphic Era Hill University, Dehradun, Uttarakhand, India*

[2]*School of Management, IMS Unison University, Dehradun, Uttarakhand, India*

[3]*Department of Management, Graphic Era Deemed to be University, Dehradun, Uttarakhand, India*

*Corresponding author

ABSTRACT

The proliferation of robo-advisors epitomizes a paradigmatic transformation within contemporary investment management, wherein algorithmic intelligence, machine learning, and data-driven architectures are supplanting conventional advisory mechanisms. These digital platforms democratize financial planning by providing cost-efficient, accessible, and personalized portfolio solutions that transcend traditional barriers of wealth thresholds and geographical limitations. Their operational efficacy stems from the amalgamation of automated asset allocation, dynamic rebalancing, tax optimization, and behavioral-bias mitigation, thereby enhancing investor engagement and rationality in decision-making. Historical antecedents such as algorithmic trading and digital brokerage systems provided the foundation for the advent of robo-advisory models, which have progressively evolved from basic questionnaire-driven tools

Robo-Advisors and Artificial Intelligence in Human Resources Management: Revolutionizing HR. Sanjay Taneja, Swati Gupta, Girish Lakhera, Mohit Kukreti, and Ercan Özen (Eds.)

DOI: 10.1201/9781779641410-13

to hybrid frameworks integrating human expertise with artificial intelligence. Despite their disruptive ascendancy, challenges remain in the form of algorithmic opacity, regulatory volatility, and market sensitivity, necessitating vigilant oversight and ethical governance. Moreover, the adoption trajectory—particularly among millennials and digitally inclined cohorts—illustrates a generational shift toward technologically mediated financial intermediation. Future prospects suggest an intensified convergence of big data analytics, predictive algorithms, and holistic wealth management ecosystems, positioning robo-advisors as pivotal instruments in shaping the global financial services landscape.

13.1 INTRODUCTION

Robo-advisors are online services which use automation, algorithms, and artificial intelligence (AI) to offer financial planning, portfolio management, or investment advice to end-users (Faloon and Scherer, 2017). The development of Robo-advisors amounts to a complete departure from traditional methods in the swiftly transforming context of contemporary investment management, marking a time of efficiency, automation, and data-centric views. The financial market has experienced a transformation as these complex financial platforms were absorbed into investment management. Their operations are based on sophisticated technology such as AI and complex algorithms (Waliszewski and Zięba-Szklarska, 2020). Robo-advisors are digital systems that help people invest their money. They have been designed using cutting-edge AI to help people determine what they want to achieve with their investments, ensuring the selection of investment goals is achieved through the making of simple goal-based investment choices. They are also programmed with algorithms that keep portfolios balanced and diversified (Sironi, 2016).

The unique ability of Robo-advisors to search through large datasets, evaluate current market trends, and create highly personalized investment portfolios based on individual financial goals and risk tolerance levels is what makes them so transformative (Boreiko and Massarotti, 2020).

Making investment services democratic is like a paradigm change, as this move eliminates all the major hurdles that have been responsible for the non-availability of well-qualified finance specialists who are in a position to give out professional help on financial matters involving money matters and investments generally. Robo-advisers destroy those obstacles by providing an affordable, readily available alternative to democratize financial advisory services while at the same time exposing strategic investment-management choices to a broader audience. This inclusion alters the mechanics of wealth generation and management fundamentally, not only because it empowers a broader range of investors, but also because it promotes financial knowledge and participation in the process.

Automated decision-making platforms optimize return on investment and conceal risk in dynamic and ever-changing market settings. Robo-advisors alter and enhance their plans instantly, something traditional methods do not frequently manage. This is done through a combination of algorithmic intelligence and machine learning (ML) technologies. Thus, investors are able to gain an extent of precision and efficacy formerly considered to be beyond the scope in traditional advisory services (Phoon and Koh, 2017).

Investors receive unique personalization levels from these platforms through advanced risk profiling questionnaires and AI-driven interfaces. Apart from that, portfolios are made in a way that corresponds with particular financial goals; this improves client engagement and satisfaction. Overall, the simplicity and openness characterizing the user experience model contribute to its friendliness in investment management. At the same time, the entire process is made easier because investors can observe everything.

On the other hand, putting Robo-Advisors into investment management has some hurdles. There are certain algorithmic biases and exposure to extreme market conditions that need to be considered. It is essential to strike a proper balance between machine-driven processes and human oversight and control to overcome these hurdles efficiently (Belanche et al., 2023). As a result, hybrid Robo models are emerging in the financial advisory space, blending the best of what Robo-Advisors have to offer with an invaluable human touch. The Robo-Advisors' involvement has significantly changed how investment monitoring is done.

Given this technological change in the financial industry, we must understand the advantages and challenges that come with Robo-Advisors. In the future of investment management, technology will be at the forefront, and Robo-advisors will lead the way.

13.2 EVOLUTION OF ROBO ADVISORY MODELS VIS-A-VIS INVESTMENT MANAGEMENT

The emergence of automated investment solutions saw the genesis of Robo-Advisor models that focused on risk management, portfolio diversification, and asset allocation as major areas. By using certain algorithms, these simplest versions were able to provide customers with personalized advice according to their time frames, goals, and tolerance to hazards. Tech-savvy consumers who wanted to invest quickly and easily found these automated platforms affordable and easy to use.

Robotic advisors these days should augment rather than threaten the financial services that human advisors provide, particularly since hybrid models have come up. The importance of hybrid models—which blend the expertise of human advisors with the technological benefits offered by Robo-advisors—has increased (Deloitte, 2016). With the help of this hybrid method, customers may take advantage of the efficiency and automation of digital platforms for investment management, while still receiving personalized advice and human connection.

A brief overview of the various Robo Advisory Models is as follows:

1. **Basic Robo Model:** In this model, consumers respond to a questionnaire and then receive offers for individual items or portfolio allocations based on listed investment products. Clients manage their own real product-based portfolios without bank or broker API integration (Deloitte, 2016).
2. **Intermediate Model:** This model has diversified investment portfolios created using a range of its funds, non-controversial investment accounts opened, and direct orders placed for them as an integral part of their services, while investment advisory services delivery is partially automatic, although they are professionally overseen manually by the investment managers (Deloitte, 2016).
3. **Hybrid Robo Advisory Model:** This model enhances both customer experience and quality of investment management services by merging human knowledge with technology. Among the model's various attributes are real-time portfolio monitoring and changes, automated service offerings, besides dynamic asset class changes based on individual's requirements, among other market conditions (Deloitte, 2016).

4. **Advanced Robo Advisory Model**: At an enhanced level of automated financial advice, this method can be described as Robo-advising, as it uses self-educating AI investment algorithms, detailed risk control measures, and profiling surveys, which culminate in direct investments (Deloitte, 2016) (Figure 13.1).

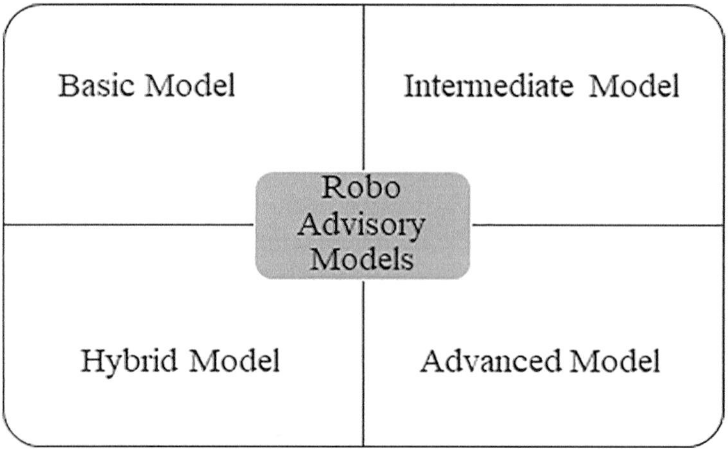

FIGURE 13.1 Robo advisory models.

Source: Compiled by Author.

13.3 HISTORICAL CHANGES IN THE INVESTMENT MANAGEMENT SECTOR

Many significant turning points in investment management have come about as a result of alterations in consumer preferences as well as technological advancements. As the financial markets have grown more intricate and linked, people have looked for creative ways to overcome these obstacles and reach their financial objectives. The rise of Robo advisors has simplified access to complex investment management solutions for individuals. Robo advisors offer personalized investment management services at significantly lower costs. This paradigm shift has empowered investors of all backgrounds to take control of their investment affairs.

Here is a detailed overview of the historical changes in the investment management sector:

1. **Evolution of Personal Financial Advisory**: Investment advisers began serving the American middle class in the 1970s, which

marked a dramatic change in the accessibility of financial advice and the beginning of the democratization of investment services. Evolution of Personal Financial Consulting: Over time, the services of asset management and personal financial counseling have changed from being pricey and exclusive to being more widely available and affordable for affluent people (Waliszewski and Zięba-Szklarska, 2020).

2. **Influence of Technological Progressions**: With the advent of the Internet in the 1990s, communication and accessibility underwent significant transformations. The development of internet trading platforms was made possible by this technological breakthrough, which expanded the pool of potential investors. The quick advancement of IT technology and the creation of better computer, smartphone, and tablet communication channels lowered transaction costs and made it simpler for investors to access information (Waliszewski and Zięba-Szklarska, 2020).

3. **Introduction of Algorithmic Trading:** The advent of Algo Trading, or algorithmic trading, marked a critical turning point in the development of investment management services. Fully automated investment instruments were made possible by algorithmic trading, opening up new avenues for prospective investors to participate in the financial markets more effectively and efficiently. Automation and algorithm tactics have provided the foundation for coming innovations in digital financial services (Waliszewski and Zięba-Szklarska, 2020).

4. **Digital Transformation and Robo-Advice**: In the 21st century, traditional investment management has been challenged by the rise of Robo-advisors. They came up by offering all-automated investment solutions to an assortment of clients that include limited-budget small investors and high-net-worth individuals. Through its cost affordability, simplicity, and access, the Robo-Advice platform has completely changed the manner in which people can access and manage their investments (Waliszewski and Zięba-Szklarska, 2020).

13.4 SERVICES OFFERED BY ROBO ADVISORS UNDER INVESTMENT MANAGEMENT

1. **Portfolio Creation**: Robo-advisors are algorithm-driven portfolio managers who create investment portfolios based on individuals' personalized risk profiles, time horizons, and investing goals. This ensures the custom-made portfolios meet each investor's specific requirements, hence using a diversified and optimized investment approach (Beketov et al., 2018).
2. **Risk Customization:** Clients who choose to come up with portfolios without human intervention can receive individual assessments of risks and customizations. Robos will, in the process of determining the appropriate risk level for each client's set of investments, make use of some details including age, income opportunities, risk aversion, as well as investment goals, while at the same time bearing in mind the different preferences clients may have had. This customization process guarantees that the investment strategy is in line with the financial goals and tastes of each customer (Jung et al., 2018).
3. **Asset Allocation**: Robo-advisors handle investments by optimizing asset allocations within customer investment portfolios. They rebalance portfolios, adjust asset allocations, and make sure they are diversified using algorithms. To maximize returns for customers with minimal risks, they keep an eye on asset performance as well as market conditions (Beketov et al., 2018).
4. **Market Adaptation**: Robo-advisors use data analytics and technology to anticipate and respond to changes in the market. These tools help them respond to economic indicators, asset price movements, and market trends as they develop. In doing so, they capitalize on any openings that may come up while at the same time reducing any possible hazards in variable market scenarios (Anshari et al., 2022).
5. **Quantitative Evaluation:** Quantitative portfolio evaluation is done by Robo-advisors, who use algorithms and automated processes for data analysis on investment performance, various risk levels, and portfolio composition. In optimizing investment

choices and ensuring effective portfolio management, this approach is quantitative and thus facilitates the usage of Robo-advisors (Jung et al., 2018).

13.5 PROCESS OF INVESTMENT MANAGEMENT SOLUTIONS OFFERED BY ROBO ADVISORS

Automated platforms simplify the investment process, from when an investor creates an account to evaluating risks. This includes filling out a form that figures out what goals you need to invest for and the likelihood that you will risk your money (Sironi, 2016). Robo-advisor algorithms construct varied portfolios based on the investor's preferences; hence, a mix of assets which include, but are not limited to, shares, debentures, and ETFs (Beketov et al., 2018). This automatic rebalancing guarantees that the investor's objectives are maintained throughout the lifetime of the plan. Indeed, a summary of how investment management works using Robo-advisors is as follows (Figure 13.2):

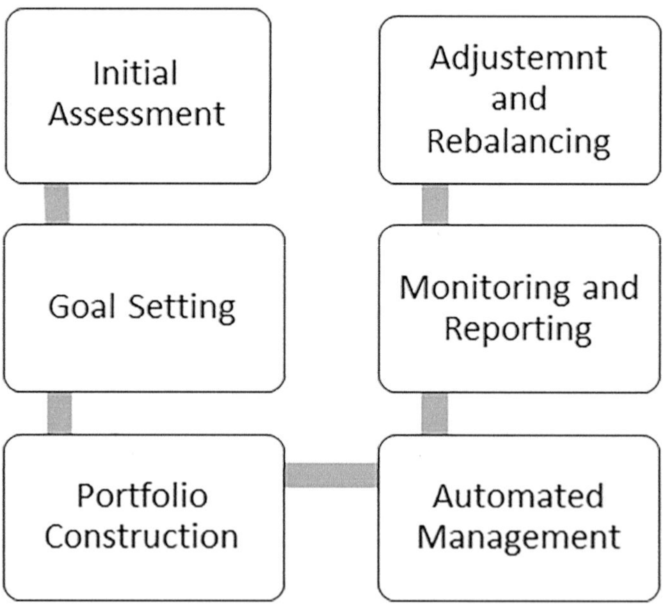

FIGURE 13.2 Robo advisor investment management process.

Source: Compiled by Author.

1. **Initial Assessment**: Initially, investors undergo an assessment to offer data about their related risk inclination, investment objectives, timing, and economic situation. This information undergoes objective analysis by Robo Advisors to ascertain the risk profile and financial product preferences that each investor might have.
2. **Goal Setting:** Investors define their short-term and long-term investment goals, like retirement planning or saving for education, and then Robo-advisors come up with tailor-made approaches to investments that would suit each goal, risk-bearing capability, time frame, and expected profits for all investors or customers.
3. **Portfolio Construction**: Portfolio construction involves figuring out how individual investors can spread their assets among different types of assets based on their exposure tolerance and end goals. Given the different categories of investments, they can seek guidance from Robo Advisors. By spreading investments over such securities as bonds or cash equivalents, as well as other belongings, diversification is put at the forefront of influencing risk effectively.
4. **Automated Management**: Robo advisors use algorithms to automate the management of portfolios by constantly checking and maintaining target asset allocations, and rebalancing them, among other things. Some Robo advisors may also provide tax optimization strategies that aim at reducing taxes while increasing returns that are left over after taxation.
5. **Monitoring and Reporting**: Investors can monitor their investment portfolios using real-time information, which allows them to observe the performance in relation to their set objectives. Robo-advisors provide reports on a frequent basis, explaining investment returns attained, charges paid, as well as regarding the financial goals and achievements.
6. **Adjustments and Rebalancing:** According to financial circumstances that have changed among the investors or market conditions, Robo Advisors could be advised to change investment strategies, such as portfolios being realigned automatically to maintain the required asset distribution.

13.6 EXPLORING MARKET PENETRATION AND GROWTH OF ROBO ADVISORS

Market penetration, Robo-advisor users, and assets under management (AUM) are some of the key metrics that give insights into the market penetration and growth of Robo-advisor services, market penetration being a percentage of active paying customers or accounts over the total market population to portray adoption and utilization of Robo-advisor platforms. The number of Robo-advisor users in this case refers to active paying customers or accounts at given periods, which point to an increase and attraction to investors in general. The size of the global Robo-advice market was forecast to be US$6.61 billion in 2023, and between 2024 and 2030, it is expected to develop at a CAGR (compound annual growth rate) of 30.5% (PR NewsWire, 2024) (Figure 13.3).

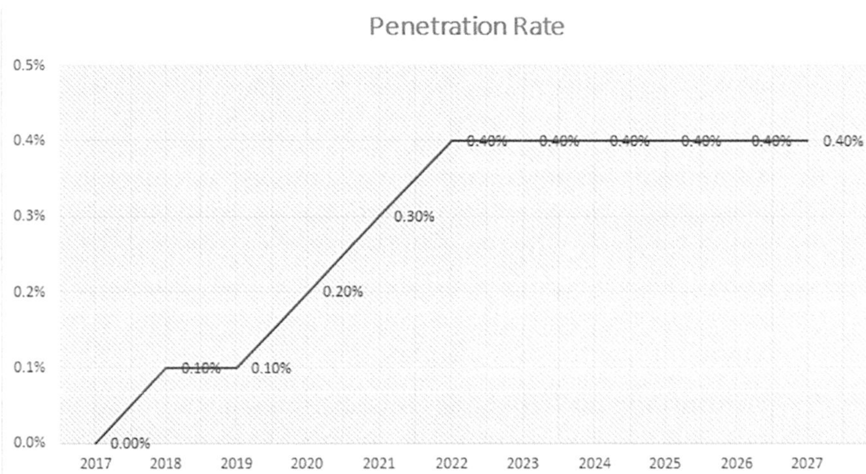

FIGURE 13.3 Penetration rate per year.

Source: Calculated by Author.

The Penetration Rate box provides insight into the proportion of active paying customers or accounts relative to the total population of the selected market, segment, or region for each year. In 2017, no active paying customers were recorded, resulting in a penetration rate of 0.00%. Over the following years, the penetration rate slowly increased, with 0.10% recorded in both 2018 and 2019, indicating a small fraction of the market engaging in paying transactions. By 2021, this proportion rose to 0.30%, suggesting a gradual uptake in paying customers. This trend of incremental growth continued through 2027, with the penetration rate consistently increasing to reach 0.40% (Statista, 2024). Although seemingly modest, these percentages represent a growing segment of the market population engaging in paid transactions, reflecting a gradual but steady expansion of active paying customers over the years (Figure 13.4).

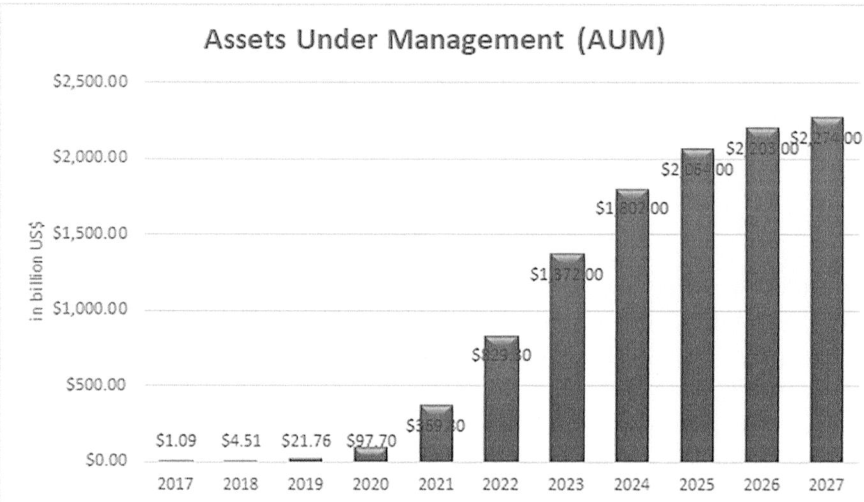

FIGURE 13.4 Assets under management forecast.

Source: Calculated by Author.

The average AUM per user exhibit a remarkable upward trend from 2017 to 2027. Beginning at $1.09 billion in 2017, the figures surged consistently each subsequent year. By 2018, assets had quadrupled to $4.51 billion, signaling early signs of growth. The momentum accelerated dramatically in 2019, with assets rising to $21.76 billion, marking a significant expansion. The trend continued its upward trajectory in 2020, with assets reaching $97.70 billion, reflecting substantial growth and investor confidence. In 2021, this growth surged even further, with assets soaring to $369.80 billion, underscoring robust market performance. There was a trend of rapid expansion through the year 2022 with more than $829.30 billion in assets, growing double compared, so that showed the optimism of investors over time.

The total AUM for 2024 and beyond is estimated, providing a reflection based on prior tendencies and market analysis. Earlier, by 2023, total properties had a value of $1372.00 billion, indicating more growth. Already by 2024, it is projected that the assets will have exceeded the $1,800 billion mark to stand at $1802.00 billion, which shows that they are on an upward trend (Statista, 2024). The report indicates that the growth trend will continue to boost up in 2025 and 2026, with assets rising to $2064.00 billion and $2203.00 billion correspondingly, which signals the sustained inflow of investment funds as well as unflinching market confidence. It is expected that by the close of 2027, total money managed would hit $2274.00 billion, indicating a mature market environment.

13.7 ADOPTION OF ROBO-ADVISORY SOLUTIONS BY INVESTORS

Investors choosing to use automatic investment plans instead of hiring traditional advisers is a significant trend that indicates over-reliance on automation in finance. "Robo-advisors" are algorithms or artificial intelligence software applications that give counsel on investing as well as invest assets for people. Studies frequently look at characteristics of users of Robo-advisors and their reasons behind their decision to go for Robo Advisory Services. Research shows that, in general, Robo-advisors are more favored by young people like millennials and Generation Z because they are comfortable with modern technology and prefer digital-first encounters over face-to-face meetings (Yi et al., 2023). Similarly, lower expenses compared to conventional financial advisor fees, and interest in economizing on expenditure, appeal Robo-advisors even to more mature customers (Figure 13.5).

The above chart displays how many millions of active users Robo-advisors have activated in this selected market each year. In 2017, this service had 2.36 million users, showing the first group of users who used the robot advisor platform. Over time, this number of users exceeded 5.04 million in 2018 and 9.83 million in 2019. As of the close of 2020, this stood at 16.53 million users before hitting 23.29 million users during 2021. These figures indicate an increasing penetration of Robo-advisor advisors among the public. Up to 34.02 million users saw a steady increase in their numbers until 2027 (Statista, 2024). This data serves as proof that, year

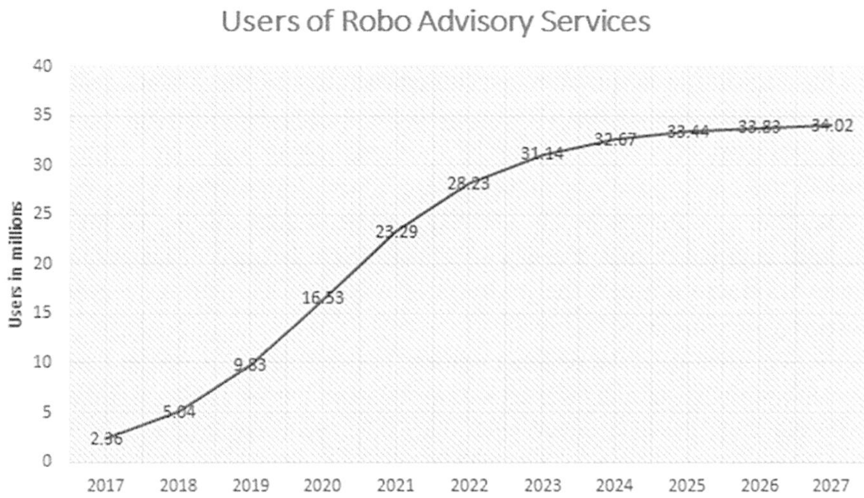

FIGURE 13.5 Users of Robo advisory services.

Source: Calculated by Author.

after year, Robo-advisors advisor platforms have been growing steadily, indicating a rise in their demand among more and more people who are looking for automatic investment alternatives.

13.8 DETERMINANTS AFFECTING THE ADOPTION OF ROBO ADVISORY SERVICES

Knowing how Robo-advisors are accepted is very important for different reasons while handling investment management. For starters, it aids in understanding the behavior of investors, providing financial institutions as well as Robo-advisor providers an opportunity to customize their packages and marketing strategies (Phoon and Koh, 2017). They can focus on investors' concerns, promote Robo-advisors' benefits, and reduce obstacles to adoption by understanding the psychological, social, and technological elements that affect adoption (Ahn et al., 2020). Familiarity with these determinants makes it easy to design and come up with Robo-advisor platforms that are favored by the investors and have the necessary standards. Robo-advisor providers can increase their market penetration and satisfaction levels among users as long as they build such characteristics as perceived usefulness, ease of use, and social acceptance in them.

In addition, a comprehensive understanding of the factors that influence the adoption of Robo Advisor technologies by financial regulators and policymakers would enable them to come up with appropriate guidelines and regulations that promote innovation, market efficiency, and protection of consumers. By identifying the factors driving or inhibiting its use, regulators may design an atmosphere that promotes responsible innovation and protects the interests of investors as well as promotes the integrity of markets.

It is critical for various stakeholders in the asset management sector to understand why people take up Robot Advisors, as this understanding will help them get the best out of such platforms, promote financial inclusion, and enhance the industry's effectiveness and competitiveness (Belanche et al., 2019).

Three main theories are being looked into as driving people to use Robo Advisors: the Theory of Planned Behavior (TPB), the Technology Acceptance Model (TAM), and the Unified Theory of Acceptance and Use of Technology (UTAUT).

13.8.1 THEORY OF PLANNED BEHAVIOR (TPB)

In 1985, the TPB was developed, based on the Theory of Reasoned Action (TRA), which was initially suggested by Ajzen and Martin Fishbein (Ajzen, 1985). TPB differs from TRA as it introduces perceived behavioral control into its structure. It is often used to explore investment decisions about the use of Robo Advisors. This psychological model examines people's actions through their attitudes, social norms, and perceived capacity for self-regulation.

1. Attitudes are how investors perceive Robo Advisors either positively or negatively.
2. Subjective norms are the effects of social factors such as peer reviews for financial decisions on investors' choice of using Robo Advisors.
3. Perceived behavioral control concerns with the investor's assessment of their capability to effectively apply Robo Advisors.

With an analysis of such constituents, TPB gives an understanding of what drives and inhibits the choices investors who want to use Robo Advisors for their money management make.

13.8.2 TECHNOLOGY ACCEPTANCE MODEL (TAM)

The TAM system was developed in 1985 (Davis, 1985). It was later developed in 1989 (Davis et al., 1989). It is a way of understanding how people accept and use technology. Though an evolution from previous models like the TRA or Planned Behavior Theory, it is notably meant for addressing Information Systems and technologies adoption by individuals' requirements. In situations where there are Robo Advisors, TAM has useful information on why people invest in them. In other words, TAM primarily considers how ordinary individuals feel about a given technological infrastructure—it encompasses both perceived usefulness and ease of use. From one's standpoint, the first criterion characterizes their ideas on whether employing Robo Advisors can be lucrative and instrumental in attaining their financial aspirations. The second aspect appreciates the user-friendliness of Robo Advisors according to prospective investors.

13.8.3 UNIFIED THEORY OF ACCEPTANCE AND USE OF TECHNOLOGY (UTAUT)

People's preference for Robo Advisors can be explained by the usefulness of UTAUT theory (Venkatesh et al., 2003). It focuses on comprehending individual intentions of using technology, along with their reactions towards acceptance and usage aspects. In relation to Robo Advisors, there are four major factors that determine adoption decisions according to UTAUT:

1. **Performance Expectancy:** It is described as investors' expectation of the benefits and effectiveness of employing a Robo Advisor to manage their investments.
2. **Effort Expectancy:** It emphasizes how easy investing through Robo Advisors can be while addressing concerns about its usage convenience for investors.

3. **Social Influence:** This reflects the social factors influencing an investor's decision-making, including recommendations given by his/her peers in the context of finance-related matters, or even industry professionals, among others.
4. **Facilitating Conditions:** It deals with investors' perception of resources, support, as well as the available infrastructure that would encourage consumers to adopt and use.

When these things are considered, UTAUT presents an elaborate structure to explain why investors accept Robo Advisors and their behavior towards this. This structure, therefore, will allow both scholars and professionals to generate ways of getting more people to take up these items and effectively use them in managing finances.

13.9 EXPLORING HOW ROBO ADVISOR FEATURES DRIVE INVESTOR ADOPTION

There are many things that motivate investors to adopt Robo-advisors, such as accessibility, cost effectiveness, and technological advancement, which are core characteristics of such platforms. Investing through automated platforms helps individuals access expert investment management services, build portfolios with many different investments, as well as attain their financial targets with more ease and speed. The investment landscape may be reshaped increasingly by Robo-advisors in the wake of technological advancements and changing investor priorities and behavior when it comes to investing, while financial planning and investment management undergo a revolution.

1. **Accessibility and Convenience**: One of the basic characteristics of Robo Advisory services is the increased availability and convenience that Robo-advisory services provide. Weak access to investment opportunities has been experienced by people with small portfolios or low financial knowledge in historical times (Gold and Kursh, 2017). Usually, people who wanted to get professional investment advice or manage their portfolios had to have a lot of money and mostly had to deal with conventional financial advisors or wealth management firms in the past. The situation, however, is different as Robo-advisors have come by.

These platforms have enabled people to invest more easily than ever before. Investment accounts can be created with ease, financial goals set, and performance of investments observed using these automated systems, even from one's laptop. This user-friendly design makes becoming a money manager or financial planner more possible than ever before and opens investment opportunities up for everyone.

2. **Cost Effectiveness**: Another significant advantage for Robo Advisory Services is that it is more affordable when compared with old-style financial advisors (Phoon and Koh, 2017). Traditional financial advisors typically collect charges that vary depending on AUM; as a result, there is a substantial amount of money for small investors and people with average portfolios. Therefore, using these new tools is much cheaper because it involves less human input, such as administration or customer service representatives, which would have otherwise increased operational expenses. Using automation and digital platforms, they provide financial services at annual percentage rates or flat fees, which remain nominal irrespective of the portfolio size. This affordability factor expands access to investing for a broader group of investors, thereby democratizing the realms of investment management and financial planning.
3. **Advanced Investment Management Solutions**: Technological advancements have mostly driven the adoption of Robo-advisors, especially in the areas of artificial intelligence, machine learning (ML), and data analytics. These innovations allow Robo-advisors to instantly analyze huge sets of data, such as investor profiles, market trends, and economic indicators, which are necessary for making data-driven investment decisions (Anshari et al., 2022). The power of algorithms and automation allows Robo-advisors to optimize portfolio allocation, rebalance assets, and minimize risk exposure more efficiently than traditional human advisors.
4. **Personalization:** Algorithms and data analysis are used by many Robo-advisors to offer personalized investment guidance that suits their clients' financial objectives, risk preference levels, and investment duration (Kwon et al., 2022). The advantage of using online platforms and mobile applications is that they allow

investors to check over their portfolio on their own time without having to meet advisors, hence making it more convenient on their part.

5. **Portfolio Diversification:** Automated processes allow Robo-advisors to distribute investments among various asset classes in a more efficient way, thus helping to mitigate the risk of becoming over-exposed to a single investment, thereby making them stress the significance of diversification in the creation of properly well-balanced investment portfolios (Beketov et al., 2018).
6. **Transparency** Investors can see how their money is being managed thanks to clear insight into investment strategies and transparent fee structures that are usually provided by Robo-advisors (Jung et al., 2018). The skepticism of traditional financial institutions is one reason that helps to build trust and bring about confidence among investors.
7. **Tax Loss Harvesting** Tax-loss harvesting: Robo-advisors use tax-loss harvesting strategies that cut down on tax liabilities for investors. This is done through automatically trading investments that are less competitive so that they mitigate any negative impacts on tax income. Through tactful realization of these losses, tragically advised robots manage to improve after-the-fact earnings to their clients, thus enabling them to achieve the most financially from their investments while at the same time reducing the taxes that they owe (Gold and Kursh, 2017).

13.10 PERFORMANCE EVALUATION OF ROBO ADVISORS

In order to appraise Robo Advisors' operations, their primary task is to see that they earn higher profits for investors based on adopted investment objectives and risks which they are prepared to undertake. Ordinarily, investors are required to answer a few questions that would help Robo-advisors' managers know how much risky behavior they can afford and invest their cash in various assets at market. One would then measure how well these portfolios have done by comparing them to market indexes or target returns that are relevant, which help us determine how effective they have been in achieving the desired objectives of the investors (Nadeem and Shaheen, 2021).

The performance of Robo Advisory Portfolios is influenced by a number of factors, such as asset allocation strategy, that defines how one's investments are spread among various categories based on such criteria as level of risk aversion or investment targets. Effective allocation strategies are aimed at controlling risks as well as enhancing profits, thus affecting the performance of portfolios directly.

The way you invest, whether it is through an active or passive strategy, impacts several variables such as expenses and risk levels that go into the final calculations for what returns are earned on this investment option. Meaning that your net gain from this opportunity might be either reduced or increased depending on its cost structure, which comprises fees and charges. In addition, regular rebalancing or after-tax-loss opportunities should also help improve returns since they align it with investor goals while minimizing taxes paid on gains realized.

Another important factor to consider is broader market conditions, economic trends, and geopolitical events that play a role in this. It also influences returns when stocks fall sharply. The economy, however, has so much volatility in it, although not all times are bad economically, as sometimes, we have a bull economy, though uncommon. Since volatility is greater for stocks than bonds generally, and falling prices cause losses to investors through bankruptcy, portfolio performance optimization requires an understanding and management of these variables. It is important to evaluate the performance of Robo-advisors in relation to investor goals and risk tolerance before considering the automated investment management options they provide, such as low management costs or diversified investment instruments. Furthermore, investors are encouraged to review their investment holdings periodically so that these can be realigned suitably according to future financial targets.

13.11 ROBO ADVISORS IMPACT ON BEHAVIORAL BIASES

Robo-advisors are a promising solution for dealing with different behavioral biases that come as a result of prospect theory, including disposition effect, home-biased portfolio trend chasing, and underperformance chasing (Lisauskiene and Darskuviene, 2021). By automating decision-making, Robo-advisors can assist investors in behaving non-irrationally and avoid other forms of bias that might lead them to make less optimal investment decisions.

Studies have proven that Robo-advisors have a meaningful impact on self-directed saving plans by enhancing diversification, enabling passive investment in ETFs, as well as being cost-efficient. Also, the use of robotic advisors through automated order execution has the potential to overcome a lack of willpower and procrastination characteristic of individuals, especially when making decisions. Though there is a dearth of literature about Robo-advisors and how they influence behavioral biases, mostly concentrating on specific ones (Bhatia et al., 2020), diverse findings point to the necessity for more inquiry so as to determine the general effects of Robo-advisors on different kinds of behavioral biases. In totality, amongst investment management assets under its purview, Robo-advisors could go a long way in curbing these biases, hence improving the decision-making capacities of personal capitalists.

13.12 TRADITIONAL INVESTMENT MANAGEMENT VS ROBO ADVISORY INVESTMENT MANAGEMENT

Robo advisors have numerous advantages over conventional advisory approaches. For one thing, they do not have high or even low minimum investment thresholds, so people can start investing with little money. Indeed, there are platforms where you can have zero savings yet still be able to register and put in very minimal starting capital. This feature enables Robo advisors to be accessible by a wide pool of investors, one of whom might have little money to afford other investment opportunities, since it aligns with the modern portfolio theory and always beats self-managed or mutual funds on a risk-adjusted basis, thereby giving them a competitive advantage over other conventional ways (Belanche et al., 2023). The former's portfolio management skills become especially valuable in the context of time; hence, they are good for long-term wealth creation. In addition, with emotion being replaced at some point by algorithms and data, Robo advisors take care of this psychological part of traditional advisory services by steering clear of emotional biases and constantly keeping the risk levels static. In addition to quantitative portfolio evaluation and management services, they offer an automated advice process lasting from customer profiling to asset allocation and strategy assessment, which ensures easily attainable investment solutions that are less costly (Gold and Kursh, 2017). Furthermore, the rise of

Robo-advisors – algorithm-based tech systems – has significantly shaken traditional investment services as they bring about automatic portfolio management procedures and changeable strategies. These allies have been able to cut down on management fees by a huge percentage and thus may be expected to hold substantial AUM due to their disruptive nature paired (Waliszewski and Zięba-Szklarska, 2020) with low-cost innovations in investments as well.

13.13 THREAT TO ROBO ADVISORS

A major danger that Robo Advisors face is from regulatory challenges and modifications in the regulation of the market. For Robo Advisors, keeping up with financial legal requirements, laws controlling the protection of client information data, and measures aimed at realizing openness is necessary in order to avoid legal ramifications, damage to one's reputation, and changes in activity regimes as a result of regulatory frameworks within the Financial Sector (Lam, 2016). Furthermore, instability in regulations that integrate companies such as Robo-advisors could tamper with market growth if any compliance-related change interfered with operations; hence, profitability and acceptability to the market is hugely hampered (Waliszewski and Zięba-Szklarska, 2020). Urging for enhanced regulation of Robo Advisors due to fears like algorithm monitoring, conflicts among algorithms and conflicts among traded algorithms, as well as personal information clients' exploitation, accentuates the necessity for regulatory mechanisms efficient in curing difficulties and hazards posed by such platforms (Baker, 2017). Without regulations, Robo Advisors might get into more scrutiny, legal challenges, and compliance issues. On the other hand, hard rules or not following current guidance could also help reduce the profitability of Robo Advisor platforms through cost increments, fines, or operational shifts needed. This shows how important it is for them to vigilantly watch out for any changes that might take place with regard to legal requirements and at the same time manage any risks so as to remain profitable over time and quite stable in the long run (Fein, 2015). Addressing these regulatory threats is essential for Robo Advisors to uphold trust with clients, regulators, and stakeholders, ensuring their viability and prosperity in the financial services industry.

13.14 FUTURE OF ROBO ADVISORS

Robo-advisors seem promising in the investment sector that manages services; trends and advancements are making it have this appearance. First of all, there is an expectation that more market penetration will occur through technological inventions, changes in demand patterns from customers who are looking for digital options, and the attractiveness of reduced costs with regard to customized investment counsel they offer. The direction the growth will take is in the adoption of investment management solutions involving digital components. Secondly, it is anticipated that future Robo-advisors would use advanced algorithms, ML, as well as data analyses with the aim of offering highly personalized services limited only by individual preferences, appetite for risk, and ultimate financial objectives (Phoon and Koh, 2017). The future augurs that the combination between artificial intelligence (AI) and big data analytics will boost Robo-advisors' power in offering advanced investment strategies, predictive analytics, and risk management solutions whose accuracy and timeliness for clients can be improved upon (Anshari et al., 2022). Robo-advisers are positioning themselves as complete financial management centers by offering more than just investment management services; they offer financial planning inclusive of estate planning, tax optimization, insurance advice, among others (Fisch et al., 2019). In building trust or confidence, transparency must be maintained, meaning creating guidelines to be followed in ensuring complete transparency, data protection, as well as adhering to financial laws are considered indispensable by many industry players (Yang et al., 2022).

13.15 CONCLUSION

Robo advisors are classified as the new powerhouses of interference in the investment management industry due to their effective and available solutions to all levels of investors. As discussed in this segment, their progression from basic algorithmic models to advanced platforms has changed the approach individuals take when it comes to investment choices. Historical changes in the investment sector and technological improvements have spurred global growth and adoption of Robo Advisors. Despite regulatory challenges and potential threats they face, they are becoming popular, especially due to their transparency, cost-effectiveness, and ability to mitigate behavioral biases. Robo advisory appears to be taking a positive direction from the look of things, as new developments in this sector have been reported that could meet diverse investors' needs. The financial industry undergoes constant changes; hence, the rising importance of Robo advisors in determining where investment management is heading. The rise of different technologies and how they are currently changing while being used in financial management ensures Robo Advisors have significant functions in adjusting to the ever-changing financial services' field.

KEYWORDS

- **algorithmic investment**
- **behavioral bias mitigation**
- **digital wealth management**
- **financial technology**
- **robo-advisors**

REFERENCES

Ahn, W., Lee, H. S., Ryou, H., & Oh, K. J. (2020). Asset allocation model for a Robo-advisor using the financial market instability index and genetic algorithms. *Sustainability, 12*(3), 849. https://doi.org/10.3390/su12030849 (accessed on 11th August 2025).

Ajzen, I. (1985). From intentions to actions: A theory of planned behavior. In J. Kuhl & J. Beckmann (Eds.), *Action Control: From Cognition to Behavior* (pp. 11–39). Springer. https://doi.org/10.1007/978-3-642-69746-3_2 (accessed on 11th August 2025).

Anshari, M., Almunawar, M. N., & Masri, M. (2022). Digital twin: Financial technology's next frontier of Robo-advisor. *Journal of Risk and Financial Management, 15*(4), 163. https://doi.org/10.3390/jrfm15040163 (accessed on 11th August 2025).

Baker, T. (2017). Regulating robo advice across the financial services industry. *SSRN Electronic Journal*. https://doi.org/10.2139/ssrn.2932189 (accessed on 11th August 2025).

Beketov, M., Lehmann, K., & Wittke, M. (2018). Robo advisors: Quantitative methods inside the robots. *Journal of Asset Management, 19*(6), 363–370. https://doi.org/10.1057/s41260-018-0092-9 (accessed on 11th August 2025).

Belanche, D., Casaló, L. V., & Flavián, C. (2019). Artificial intelligence in FinTech: Understanding Robo-advisors adoption among customers. *Industrial Management & Data Systems, 119*(7), 1411–1430. https://doi.org/10.1108/IMDS-08-2018-0368 (accessed on 11th August 2025).

Belanche, D., Casaló, L. V., Flavián, M., & Loureiro, S. M. C. (2023). Benefit versus risk: A behavioral model for using Robo-advisors. *The Service Industries Journal*, 1–28. https://doi.org/10.1080/02642069.2023.2176485 (accessed on 11th August 2025).

Bhatia, A., Chandani, A., & Chhateja, J. (2020). Robo advisory and its potential in addressing the behavioral biases of investors—A qualitative study in Indian context. *Journal of Behavioral and Experimental Finance, 25*, 100281. https://doi.org/10.1016/j.jbef.2020.100281 (accessed on 11th August 2025).

Boreiko, D., & Massarotti, F. (2020). How risk profiles of investors affect Robo-advised portfolios. *Frontiers in Artificial Intelligence, 3*, 60. https://doi.org/10.3389/frai.2020.00060 (accessed on 11th August 2025).

Davis, F. D. (1985). A technology acceptance model for empirically testing new end-user information systems: Theory and results [Doctoral dissertation, Massachusetts Institute of Technology]. https://dspace.mit.edu/bitstream/handle/1721.1/15192/14927137-MIT.pdf (accessed on 11th August 2025).

Davis, F. D., Bagozzi, R. P., & Warshaw, P. R. (1989). User acceptance of computer technology: A comparison of two theoretical models. *Management Science, 35*(8), 982–1003. https://doi.org/10.1287/mnsc.35.8.982 (accessed on 11th August 2025).

Deloitte. (2016). *The Expansion of Robo-Advisory in Wealth Management.*

Faloon, M., & Scherer, B. (2017). *Individualization of Robo-Advice.*

Fein, M. L. (2015). Robo-advisors: A closer look. *SSRN Electronic Journal.* https://doi.org/10.2139/ssrn.2658701 (accessed on 11th August 2025).

Fisch, J. E., Labouré, M., & Turner, J. A. (2019). The emergence of the robo-advisor. In J. E. Fisch, M. Labouré, & J. A. Turner, *The Disruptive Impact of FinTech on Retirement Systems* (pp. 13–37). Oxford University Press. https://doi.org/10.1093/oso/9780198845553.003.0002 (accessed on 11th August 2025).

Gold, N. A., & Kursh, S. R. (2017). Counterrevolutionaries in the financial services industry: Teaching disruption–A case study of roboadvisors and incumbent responses. *Business Education Innovation Journal, 9*(1), 139–146. http://elmstreetpress.com/images/V9N1_final.pdf#page=139 (accessed on 11th August 2025).

Jung, D., Dorner, V., Weinhardt, C., & Pusmaz, H. (2018). Designing a robo-advisor for risk-averse, low-budget consumers. *Electronic Markets, 28*(3), 367–380. https://doi.org/10.1007/s12525-017-0279-9 (accessed on 11th August 2025).

Kwon, D., Jeong, P., & Chung, D. (2022). An empirical study of factors influencing the intention to use robo-advisors. *Journal of Information & Knowledge Management, 21*(03), 2250039. https://doi.org/10.1142/S0219649222500393 (accessed on 11th August 2025).

Lam, J. W. (2016). *Robo-Advisors: A Portfolio Management Perspective.*

Lisauskiene, N., & Darskuviene, V. (2021). Linking the robo-advisors phenomenon and behavioral biases in investment management: An interdisciplinary literature review and research agenda. *Organizations and Markets in Emerging Economies, 12*(2), 459–477. https://doi.org/10.15388/omee.2021.12.65 (accessed on 11th August 2025).

Nadeem, R. M., & Shaheen, R. (2021). The impact of using robo-advisors on financial institutions performance.

Phoon, K., & Koh, F. (2017). Robo-advisors and wealth management. *The Journal of Alternative Investments, 20*(3), 79–94. https://doi.org/10.3905/jai.2018.20.3.079 (accessed on 11th August 2025).

PR NewsWire. (2024). Robo advisory market projected to skyrocket to USD 41.83 billion by 2030 with noteworthy CAGR of 30.5%.

Sironi, P. (Ed.). (2016). Front matter. In *FinTech Innovation* (1st ed.). Wiley. https://doi.org/10.1002/9781119227205.fmatter (accessed on 11th August 2025).

Statista. (2024). Robo-advisors—Worldwide | Statista market forecast. Statista. https://www.statista.com/outlook/fmo/wealth-management/digital-investment/robo-advisors/worldwide (accessed on 11th August 2025).

Venkatesh, V., Morris, M. G., Davis, G. B., & Davis, F. D. (2003). User acceptance of information technology: Toward a unified view. *MIS Quarterly, 27*(3), 425. https://doi.org/10.2307/30036540 (accessed on 11th August 2025).

Waliszewski, K., & Zięba-Szklarska, M. (2020). Robo-advisors as automated personal financial planners–SWOT analysis. *Finanse i Prawo Finansowe, 3*(27), 155–173. https://doi.org/10.18778/2391-6478.3.27.09 (accessed on 11th August 2025).

Yang, B., Yu, H., & Chen, J. (2022). Understanding the factors influencing adoption of robo-advisor: Integration model of TAM and TTF. *Academy of Management Proceedings, 2022*(1), 13712. https://doi.org/10.5465/AMBPP.2022.13712abstract (accessed on 11th August 2025).

Yi, T. Z., Rom, N. A. M., Hassan, N. Md., Samsurijan, M. S., & Ebekozien, A. (2023). The adoption of robo-advisory among millennials in the 21st century: Trust, usability and knowledge perception. *Sustainability, 15*(7), 6016. https://doi.org/10.3390/su15076016 (accessed on 11th August 2025).

CHAPTER 14

The Transformative Impact of Robo-Advisors on Human Resource Management: A Comprehensive Analysis

NAVJOT S. MIGLANI, NILESH ARORA,* and POONAM ARORA

University School of Business, Chandigarh University, Gharuan, Mohali, Punjab, India

*Corresponding author

ABSTRACT

This chapter explores the transformative implications of robo-advisors in Human Resource Management (HRM), underscoring how artificial intelligence (AI) and machine learning are redefining recruitment, employee engagement, performance evaluation, and strategic workforce planning. Drawing upon a mixed-methods case study from a private Indian university, the analysis reveals how AI-driven systems—particularly Applicant Tracking Systems (ATS)—streamline job postings, automate candidate shortlisting, and substantially enhance offer fulfilment rates. The findings demonstrate a 132% increase in applications, a 44% reduction in manual screening, and a 66% improvement in successful hires post-ATS adoption, thereby validating the operational efficacy of robo-advisory integration. Beyond recruitment, robo-advisors augment employee engagement through real-time sentiment analysis, facilitate personalized development trajectories, and enable predictive workforce analytics for long-term planning. Nonetheless, the study acknowledges significant challenges, including algorithmic bias, data privacy vulnerabilities, and

Robo-Advisors and Artificial Intelligence in Human Resources Management: Revolutionizing HR.
Sanjay Taneja, Swati Gupta, Girish Lakhera, Mohit Kukreti, and Ercan Özen (Eds.)

DOI: 10.1201/9781779641410-14

ethical dilemmas concerning transparency and accountability. The chapter concludes that sustainable adoption necessitates a balanced human-AI synergy, strong governance frameworks, and continuous upskilling of HR professionals. Ultimately, robo-advisors emerge as catalysts for agile, data-driven, and strategically aligned HR ecosystems, with the potential to enhance organizational performance while simultaneously reshaping the future of workforce management.

14.1 INTRODUCTION

Human resource management (HRM) is one of the many corporate domains where old operating paradigms have been transformed by the incorporation of artificial intelligence (AI). A major advancement in the way businesses manage their staff is the introduction of Robo-advisors, AI-powered technologies that automate and optimize HR procedures. Advanced algorithms and machine learning (ML) are employed by these solutions to deliver data-driven insights and optimize HR operations, encompassing hiring, employee engagement, performance management, and strategic workforce planning.

Contemporary studies have investigated the connection between HRM and innovation, with a specific focus on utilizing innovation to improve the efficiency and satisfaction derived from the recruitment process (Kaplan, 2019; Singh and Arora, 2024). Mechanical innovations that assist recruiters have become increasingly common, leading to a more digitized interface and a potential reduction in personalized interactions. It has been suggested that foresight studies and AI could be used to enhance recruitment interactions, forecasting an expanding role for AI in HRM. AI spans various disciplines such as robotics, bots, and programming, providing potential applications across the sciences, semantics, and philosophy (Clark, 2020). First introduced in 1956, AI has been in the spotlight since World War II, with its capabilities evolving to allow computers to execute tasks that require human-like intelligence. AI's objective is to enable machines to perform a wide range of tasks independently.

In today's economic environment, companies are under more and more pressure to increase productivity, cut expenses, and make better judgments. These expectations are met by the use of Robo-advisors in HRM, which automate repetitive processes, offer real-time insights, and enable more strategic HR management. This change increases operational effectiveness while freeing up HR staff members to concentrate on more strategic and valuable tasks.

ML involves the study of computer algorithms that enhance their performance through experience. As a subset of AI, it concentrates on how computers comprehend data and gain knowledge. ML allows computers to recognize patterns and make predictions, boosting the efficiency of AI systems without the need for continuous manual programming. Contrary to traditional programming where humans write code to dictate every action, ML enables AI to learn from data and adjust its behavior accordingly. This capacity to learn and improve over time makes ML a potent tool in various fields, contributing to the creation of intelligent machines capable of executing tasks independently (Arora et al., 2022).

Although the integration of Robo-advisors into HRM holds great potential, it comes with its own complications. Careful consideration must be given to matters like algorithmic bias, data privacy, and the moral ramifications of AI-driven decision-making. The goal of this study is to present a thorough examination of the effects of Robo-advisors on HRM, looking at both the advantages and disadvantages and providing predictions for these technologies' future developments.

14.2 RESEARCH OBJECTIVES

This study seeks to evaluate the advantages of AI in the realm of HRM, its implications for HR management, and the potential hurdles that may arise when incorporating AI into HR management.

The objectives of the study are as follows:

1. To comprehend the influence of artificial AI and its application to HRM.
2. To investigate how AI can aid in various HRM functions.
3. To recognize and comprehend the challenges linked with the incorporation of artificial intelligence in human resources (HR) management.
4. To develop pertinent suggestions based on the findings of the research and draw conclusions from the research evaluation.

14.3 LITERATURE REVIEW

The abundance of research on AI in HRM reflects the increasing interest in the ways that new technologies can revolutionize HR operations. The effects of AI on hiring, employee engagement, performance management, and strategic workforce planning are important areas of attention.

1. **Talent Acquisition and Recruitment:** Bersin's (2019) research on recruiting and talent acquisition emphasizes how AI may effectively mitigate biases in the hiring process. AI-powered technologies are able to evaluate applications, match applicants with positions, and forecast an applicant's performance based on past performance. These features boost the caliber of hires as well as the effectiveness of the hiring process.
2. **Employee Engagement:** According to Brown and Reilly (2020) and Arora et al. (2024), AI solutions that monitor engagement levels and offer tailored feedback can improve worker engagement. These solutions' sentiment analysis enables HR professionals to quickly detect and resolve problems, resulting in a more engaged and effective workforce.
3. **Performance Management:** The use of AI in performance management is examined by Kaplan (2021), who highlights the technologies' capacity to offer data-driven insights into worker performance. In order to promote more efficient performance management systems, these tools track key performance indicators (KPIs), spot patterns, and offer useful feedback.
4. **Strategic Workforce Planning:** In 2020, Lee and Chen investigate how AI affects strategic labor planning. Through the analysis of industry advancements, skill requirements, and employee turnover trends, AI-driven systems can forecast future labor needs. By taking a proactive stance, businesses can create strategic plans that complement their long-term objectives.
5. **Ethical and Practical Challenges:** Despite the possible advantages, the research also identifies a number of difficulties in integrating AI into HRM. The ethical ramifications of artificial intelligence (AI) are examined by Floridi et al. (2018), along with concerns about accountability, transparency, and the possibility of

algorithmic biases. Adherence to ethical AI standards and a strong governance framework are necessary to tackle these difficulties.

6. **Future Directions:** Research indicates that developments in AI and ML will have a significant impact on HRM in the future. Wilson and Daugherty (2018) suggest that optimizing the advantages of Robo-advisors would require a balanced strategy that combines human judgment with AI-driven insights. This methodology not only improves decision-making but also guarantees the moral application of AI in HRM.

14.4 METHODOLOGY

This chapter employs a mixed-methods approach, combining quantitative data analysis with a qualitative case study. Data was collected from the education sector—a private university in India that has implemented Robo-advisors in its HR department. Lately, the university has seen tremendous growth, resulting in an increase in its HR operations; therefore, utilization of AI has helped them in optimizing and streamlining its operations. The aim was to study the extent of the impact on HR practices and operations, which has been supplemented by interviews with HR professionals. The quantitative component includes statistical analysis of HR metrics pre- and post-implementation of Robo-advisors, while the qualitative component explores user experiences and perceptions.

14.5 FINDINGS AND IMPLICATIONS

➢ **Analysis of the Case Study:**

This case study compares important indicators from the pre- and post-Applicant Tracking System (ATS) implementation periods to assess the notable effects of a private university's adoption of an AI tool on the hiring process.

1. **Methodology:** Data was collected from two distinct periods:

   i. **Pre-ATS (May 2022 – April 2023):** Jobs were posted sporadically on LinkedIn (LI) or Naukri.

ii. **Post-ATS (Since May 2023):** Jobs were posted simultaneously on LinkedIn, Naukri, and Indeed using the ATS.

2. **Results:** This comparative summary evaluates the impact of implementing an ATS at a private University by contrasting key recruitment metrics from the periods before and after ATS adoption. During the pre-ATS period from May 2022 to April 2023, the university posted over 300 jobs sporadically on LinkedIn or Naukri, receiving over 20,000 applications. In contrast, since adopting the ATS in May 2023, the university posted 300 jobs simultaneously on LinkedIn, Naukri, and Indeed, resulting in a dramatic increase in applications to over 46,500, marking a 132.5% rise. Previously, there was no automated selection process, and HR manually shortlisted over 5000 candidates. Post-ATS implementation, the system automated the initial selection, deeming over 4000 candidates fit, which reduced the manual HR shortlisting workload to 2800+, a 44% reduction. Final shortlisting numbers increased by 41.7%, from over 1200 candidates pre-ATS to 1700+ candidates post-ATS, while offer releases saw a 60% rise from over 1000 manual releases to over 1600 automated releases. The number of candidates joining the university increased from over 900 to over 1500, a 66.7% increase, and the fulfilment rate improved from 75% to 94%, a 25.3% enhancement. The analysis reveals that ATS implementation significantly streamlined the recruitment process, enhancing job posting efficiency, application volume, candidate selection, and shortlisting processes, leading to higher joinings and an improved offer fulfilment rate. This study underscores the transformative potential of integrating advanced technology into recruitment strategies to achieve greater operational efficiency and effectiveness.

3. **Enhanced Employee Engagement:** A key factor in determining an organization's success is employee engagement. By monitoring engagement levels, analyzing the mood of employee comments, and making tailored intervention recommendations, robot advisors improve employee engagement. These solutions' real-time insights help HR professionals handle problems quickly, which promotes a more engaged and effective workforce. According to a study by Smith and colleagues from 2021, companies that used AI for engagement saw a 25% improvement in employee satisfaction ratings.

According to the case study, the University's use of the ATS has had a significant influence on employee engagement, mainly because it has increased the effectiveness and efficacy of the hiring process. The HR team's manual burden has decreased as a result of the streamlined and automated procedures, freeing them up to concentrate more on strategic planning as opposed to menial administrative responsibilities. Given that they can now devote more time to meaningful contacts with candidates and other value-adding activities, HR staff members' job satisfaction and engagement have probably grown as a result of this transition.

In addition, the higher fulfillment rate of 94% compared to 75% suggests that the university is performing better in swiftly and efficiently selecting qualified applicants. The entire university workforce benefits from this efficiency, in addition to the HR department. Departments can maintain consistent production and morale by experiencing less disturbance due to the faster and more accurate filling of roles.

Moreover, the increased number of applications and the improved shortlisting procedure imply that the institution is drawing and acquiring talent that is more compatible with its requirements and culture. Employee engagement usually rises when employees believe they are a good fit for both the organization and their roles. There's a good chance that recent hires will bring additional enthusiasm and dedication to the university.

To summarize, the AI has enhanced employee engagement at the university by optimizing HR processes, reducing manual workload, improving the quality of hires, and ensuring faster, more accurate fulfillment of job vacancies. This holistic improvement in the recruitment process has fostered a more positive and productive work environment.

14.6 APPLICATION OF ARTIFICIAL INTELLIGENCE IN HUMAN RESOURCE MANAGEMENT

The impact of AI on HRM is growing at a swift pace, with the potential to transform HR practices through in-depth analysis of diverse functions. Crucial areas such as hiring and selection, induction, performance evaluation, employee involvement, and retention are now supported by digital assistants. The advent of Human Resource Information Systems (HRIS) has paved the way for the incorporation of AI tools. HRIS acts as a system for collecting, storing, managing, retrieving, and validating data crucial to an organization's HR, personnel operations, and organizational unit features (Garima et al., 2020).

1. **AI in Recruitment and Selection:** Robotic advisors have made the hiring process much more efficient. They use algorithms to examine large databases and automate processes like candidate matching, resume screening, and scheduling interviews. AI-driven systems, for example, can score candidates, compare resumes to job requirements, and forecast success based on past performance. By reducing human biases, this minimizes time and expense while also improving the quality of hires (Johnson et al., 2022).
2. **Performance Management and Development:** The incorporation of Robo-advisors into performance management systems yields significant benefits. These technologies track KPIs, spot patterns, give useful feedback, and provide data-driven insights into employee performance. Furthermore, by offering training courses and professional development possibilities based on each user's unique requirements, Robo-advisors assist individualized growth plans. This strategy matches staff development with company objectives while simultaneously improving individual performance (Garcia et al., 2023; Rana and Arora, 2022).
3. **Strategic Workforce Planning:** Planning the workforce strategically is crucial to the expansion and sustainability of a firm. By forecasting future labor demands via the examination of patterns and trends in employee attrition, skill needs, and business advancements, Robo-advisors improve this procedure. AI solutions help HR departments create proactive plans that guarantee businesses are ready for any obstacles or possibilities that may arise in the future (Lee and Chen, 2020).

14.7 CHALLENGES AND ETHICAL CONSIDERATIONS

The incorporation of Robo-advisors into HRM poses various obstacles, even with their advantages. These include possible algorithmic biases, worries about data privacy, and the requirement for ongoing maintenance and updating of AI systems. To guarantee the responsible application of AI in HRM, ethical factors including accountability, transparency, and the effect on employee trust must also be considered. According to Floridi et al. (2018), addressing these issues calls for a strong governance framework and commitment to moral AI norms.

14.8 FUTURE IMPLICATIONS

ML and AI developments will have a greater influence on HRM in the future. Robo-advisors will play a bigger part in HR as they grow more complex, providing deeper insights and more complete solutions. Companies need to keep up with these advancements and make training investments for HR staff to use AI tools efficiently. To optimize the advantages of Robo-advisors and guarantee ethical use, a well-rounded strategy integrating human judgment with AI-driven insights will be essential (Wilson and Daugherty, 2018).

14.9 CONCLUSION

Robo-advisors are poised to revolutionize HRM by enhancing efficiency, improving decision-making, and supporting strategic planning. While challenges and ethical considerations must be addressed, the potential benefits of integrating AI into HRM are substantial. By embracing these technologies, organizations can create more agile, data-driven HR departments that are better equipped to meet the evolving needs of their workforce.

KEYWORDS

- **employee engagement**
- **human resource management**
- **recruitment automation**
- **robo-advisors**
- **strategic workforce planning**

REFERENCES

Arora, N., Prashar, S., Vijay, T. S., & Parsad, C. (2022). A consumer typology based on celebrity endorsement factors. *FIIB Business Review*. https://doi.org/10.1177/23197145221112749 (accessed on 11th August 2025).

Arora, N., Rana, S., Amin, R., & Tat, H. H. (2024). Robo-advisor adoption dynamics with extended technology acceptance model. In *Robo-Advisors in Management* (pp. 16–34). IGI Global.

Bersin, J. (2019). AI in HR: Challenges and opportunities. *Journal of Human Resource Management, 58*(4), 123–140.

Bhardwaj, G., Singh, S. V., & Kumar, V. (2020). An empirical study of artificial intelligence and its impact on human resource functions. In *2020 International Conference on Computation, Automation and Knowledge Management (ICCAKM)* (pp. 1–6). IEEE. https://doi.org/10.1109/iccakm46823.2020.9051544 (accessed on 11th August 2025).

Brown, A., & Reilly, P. (2020). Enhancing employee engagement through AI. *Human Resource Development Quarterly, 31*(2), 109–130.

Clark, S. (2020). How artificial intelligence and chatbots are improving employee experience. *Reworked.* https://www.reworked.co/employee-experience/5-ways-chatbots-improve-employee-experience/ (accessed on 11th August 2025).

Floridi, L., Cowls, J., Beltrametti, M., Chatila, R., Chazerand, P., Dignum, V., & Vayena, E. (2018). AI4People—An ethical framework for a good AI society: Opportunities, risks, principles, and recommendations. *Minds and Machines, 28*(4), 689–707.

Garcia, L., Johnson, H., & Smith, T. (2023). Personalized development plans through AI: A new approach. *International Journal of Human Resource Studies, 12*(1), 56–74.

Johnson, M., Lee, S., & Walker, R. (2022). Reducing bias in recruitment with AI: A case study. *Journal of Business Research, 145*, 78–89.

Kaplan, A. (2021). Optimizing performance management systems with AI. *Management Review Quarterly, 71*(3), 421–435.

Kaplan, A., & Haenlein, M. (2019). Siri, Siri, in my hand: Who's the fairest in the land? On the interpretations, illustrations, and implications of artificial intelligence. *Business Horizons, 62*(1), 15–25. https://doi.org/10.1016/j.bushor.2018.08.004 (accessed on 11th August 2025).

Lee, J., & Chen, H. (2020). Strategic workforce planning in the age of AI. *Human Resource Planning Journal, 43*(3), 211–228.

Rana, M., & Arora, N. (2022). Decoding the social media advertising influence on consumer attitude and intention. *Vision.* https://doi.org/10.1177/09722629221099588 (accessed on 11th August 2025).

Singh, M., & Arora, N. (2024). Human resource management insights on social robots' integration in the workplace. In *Robo-Advisors in Management* (pp. 104–126). IGI Global.

Smith, J., Taylor, A., & Williams, K. (2021). The impact of AI on employee engagement: A quantitative study. *Journal of Organizational Behavior, 42*(5), 712–729.

Wilson, H. J., & Daugherty, P. R. (2018). Collaborative intelligence: Humans and AI are joining forces. *Harvard Business Review, 96*(4), 114–123.

CHAPTER 15

AI and the Human Element in HR Management

SAMI ULLAH,[1,*] MOHIT KUKRETI,[2] and
MUHAMMAD REHAN SHAUKAT[3]

[1]*Faculty of Management Sciences, University of Central Punjab, Lahore, Punjab, Pakistan*

[2]*Department of Business Studies, University of Technology and Applied Sciences (UTAS), College of Applied Sciences, Ibri, Oman*

[3]*University of Technology and Applied Sciences, College of Applied Sciences, Ibri, Oman*

*Corresponding author

ABSTRACT

This chapter investigates the evolving dynamics of robo-advisors within financial ecosystems, emphasizing their disruptive potential in democratizing investment access and enhancing decision-making efficiency. By integrating algorithmic intelligence, big data, and behavioral finance principles, robo-advisors provide personalized asset allocation, systematic rebalancing, and automated portfolio optimization, thus reducing human bias and cost asymmetries inherent in traditional advisory systems. The chapter situates robo-advisory adoption within the broader framework of FinTech transformation, highlighting its implications for inclusivity, transparency, and sustainable financial management. It also addresses pressing challenges such as algorithmic opacity, data security, regulatory compliance, and ethical accountability. The discourse underscores the generational shift in investment behavior, particularly among millennials

Robo-Advisors and Artificial Intelligence in Human Resources Management: Revolutionizing HR.
Sanjay Taneja, Swati Gupta, Girish Lakhera, Mohit Kukreti, and Ercan Özen (Eds.)

DOI: 10.1201/9781779641410-15

and digitally literate investors, who exhibit higher receptivity toward technology-driven solutions. Furthermore, the narrative projects future trajectories wherein artificial intelligence, predictive analytics, and hybrid human-digital advisory models converge to create more adaptive, resilient, and ethically oriented financial services. Ultimately, the study positions robo-advisors as pivotal instruments for reconfiguring wealth management by fostering financial literacy, scalability, and global accessibility, while calling for robust regulatory and governance frameworks to ensure trust and long-term sustainability.

15.1 INTRODUCTION

Artificial Intelligence (AI) stands as a transformative force reshaping industries across the globe. At its core, AI refers to developing computer systems capable of performing tasks that typically require human intelligence. It encompasses a spectrum of capabilities, from basic pattern recognition to complex decision-making processes (Dwivedi et al., 2021). For instance, facial recognition technology deployed by social media platforms like Facebook uses AI algorithms to identify and tag individuals in photos. Similarly, autonomous vehicles rely on AI-based image recognition to detect pedestrians, traffic signs, and obstacles on the road. On the other hand, AI-driven speech recognition systems like Apple's Siri, Amazon's Alexa, and Google Assistant employ natural language processing (NLP) algorithms to understand and respond to spoken commands. These systems recognize speech patterns, decipher language nuances, and execute tasks based on user input.

In recent years, AI has found substantial application in Human Resource Management (HRM), fundamentally altering how organizations recruit, manage, and develop their workforce (Chowdhury et al., 2023). In Pakistan, one of the primary applications of AI in HRM lies in talent acquisition and recruitment (Rehman et al., 2022). Telenor Pakistan has replaced the traditional methods relying on time-consuming processes by implementing AI-powered recruitment platforms. AI-powered tools streamline HR tasks by leveraging NLP and machine learning (ML) algorithms to analyze resumes swiftly, assess candidate qualifications, and identify top prospects (Garg et al., 2022).

Moreover, AI-driven chatbots and virtual assistants are revolutionizing the employee experience (Pillai et al., 2023). These intelligent systems assist in handling routine inquiries, onboarding procedures, and providing real-time support to employees. Petroleum Development Oman (PDO) has employed AI-driven chatbots to assist employees with HR-related queries, onboarding processes, and training support. IBM's AI-powered HR assistant, "Mya," interacts with candidates, answers queries, and guides them through the hiring process, significantly enhancing efficiency and user experience while reducing the workload on HR personnel. Omantel has adopted AI-enabled sentiment analysis tools to gauge employee sentiment and satisfaction by analyzing feedback from various sources. It helps identify areas for improvement and implement targeted interventions to enhance employee morale and productivity.

However, the integration of AI in HRM is not without challenges and ethical considerations (Rodgers et al., 2023). Concerns regarding data privacy, algorithmic bias, and the ethical use of AI in decision-making processes remain pertinent. Organizations must ensure transparency, fairness, and accountability in AI-driven HR practices to mitigate these challenges and build employee trust.

15.2 THE EVOLUTION OF AI IN HRM: A HISTORICAL PERSPECTIVE

AI integration in HRM has evolved significantly over time, reshaping how organizations manage their workforce and HR functions. Tracing the historical development of AI in HRM unveils a timeline marked by key milestones and technological advancements, each contributing to the transformative role of AI in HR practices.

15.2.1 1960S TO 1980S – EMERGENCE AND EARLY APPLICATIONS

The origins of AI in HRM can be traced back to the 1960s when early computer systems began to assist with basic HR tasks like payroll and record-keeping (Abbass, 2019). However, it was not until the late 1970s and early 1980s that AI technologies, such as expert and rule-based systems, entered HR functions (Strohmeier and Piazza, 2015). The United States was at the forefront of early AI research and development during this period. Institutions like MIT, Stanford, and IBM played pivotal roles in advancing AI technologies. In the 1980s, IBM's SHRLDU (Selective

Hiring and Recruitment Linguistic Database Utility) was a pioneering AI-based system used by the HRM department for job matching. It employed linguistic analysis to match candidate resumes with job descriptions. HR professionals utilized SHRLDU to sift through large volumes of resumes, extracting relevant skills and qualifications from textual data. The system facilitated a more efficient and accurate initial screening process, enabling HR managers to focus on the most suitable candidates. These systems primarily focused on basic decision support, aiding in job matching and career guidance.

15.2.2 1990S TO EARLY 2000S – EXPANSION OF AI APPLICATIONS

The United States continued to lead in AI research and innovation. Tech giants like Microsoft, Oracle, and IBM expanded their AI capabilities and started integrating AI into HRM tools and systems. Other countries like Japan, the United Kingdom, and Germany also made notable contributions to AI development in HRM, albeit slightly less than the U.S. During this period, AI in HRM expanded its applications, leveraging ML and data analytics advancements. HR systems integrated AI-driven algorithms to automate candidate sourcing, screening, and initial assessments, streamlining recruitment processes and improving candidate selection accuracy (Rodgers et al., 2023). Financial institutions started implementing AI-powered fraud detection systems to analyze transactional data and identify patterns indicative of fraudulent activities. Companies like Fair Isaac Corporation (FICO) developed AI-based algorithms for credit scoring and risk assessment, improving accuracy in evaluating creditworthiness.

15.2.3 MID-2000S TO PRESENT – AI-POWERED HR TRANSFORMATION

The mid-2000s witnessed a significant shift as AI technologies, including natural language processing (NLP), sentiment analysis, and predictive analytics, became more sophisticated and accessible (Vrontis et al., 2022). These advancements enabled AI to play a pivotal role in various HRM domains, such as employee engagement, performance management, and workforce planning. Unilever implemented AI-driven tools for talent assessment and predictive analytics within its HRM department. These tools were used to analyze employee data comprehensively. HR managers leveraged the insights generated by these AI systems to forecast potential skill gaps, predict turnover rates, and identify areas requiring talent development initiatives (Allal-Chérif et al., 2021; Pillai and Sivathanu, 2020). This enabled Unilever's HR department to address workforce challenges proactively, retain top talent, and align workforce strategies with organizational goals.

The United States remains a key player in AI advancements in HRM, with companies like Google, Amazon, and LinkedIn pioneering the integration of AI into HR functions. However, other countries have significantly caught up and made substantial contributions. For example, China has emerged as a major force in AI development, investing heavily in AI research and development. Companies like Tencent, Alibaba, and Baidu have been actively involved in AI-driven HRM solutions, particularly in talent acquisition and management. India has witnessed considerable growth in AI technology and its application in HRM. Companies like Infosys, TCS, and Wipro have utilized AI-powered tools in recruitment, employee engagement, and performance management. European countries, including the UK, Germany, France, and the Nordic countries, have embraced AI in HRM. Various organizations in these countries have adopted AI-driven tools for recruitment, workforce planning, and employee experience enhancement. Singapore, known for its focus on technology and innovation, has seen significant adoption of AI in HRM. Companies in Singapore have been leveraging AI-powered solutions for talent acquisition, employee engagement, and predictive analytics in HR.

15.2.4 RECENT DEVELOPMENTS AND FUTURE TRENDS

The historical development of AI in HRM has progressed from basic rule-based systems to sophisticated AI-driven solutions across various HR functions (Chowdhury et al., 2023; Rodgers et al., 2023). Current trends suggest a continued evolution of AI in HRM, focusing on ethical AI, explainable AI, and integrating AI with emerging technologies like augmented reality (AR) and virtual reality (VR) for immersive HR experiences (Budhwar et al., 2023). Organizations are increasingly exploring AI-powered chatbots, VR-based training simulations, and AI-driven sentiment analysis for employee well-being (Pandey et al., 2023). Accenture is utilizing AI-powered chatbots to offer personalized assistance to employees, addressing queries related to benefits, policies, and training, thereby enhancing the overall employee experience. Oman Air, the national airline of Oman, deployed AI-driven chatbots within its HR department to enhance employee engagement. These chatbots provided personalized assistance to employees, addressing their HR-related queries, facilitating onboarding processes, and delivering training support.

15.3 AI AND ORGANIZATIONAL CULTURE

15.3.1 EMPLOYEE ENGAGEMENT

AI has profoundly transformed employee engagement by revolutionizing communication, collaboration, and personalized development opportunities (Marion and Fixson, 2021). AI-powered technologies such as chatbots, virtual assistants, and personalized learning platforms have empowered employees to collaborate seamlessly across distances while accessing tailored training and development programs. This shift towards personalization has significantly improved engagement, enabling organizations to understand and cater to individual needs and preferences (Chen, 2023).

The future of AI in the workplace is poised to bring even more significant changes. Expectations include an emphasis on even greater personalization, with AI-driven systems customizing benefits, learning opportunities, and engagement initiatives at an individual level (Dwivedi et al., 2021). Additionally, as remote and hybrid work models evolve, AI will maintain connectivity and foster a cohesive culture among geographically dispersed teams (Pianese et al., 2023). Moreover, the ethical implications surrounding AI usage will become increasingly critical, requiring organizations to address data privacy, bias, and fairness concerns (Munoko et al., 2020). Furthermore, AI is predicted to focus more on monitoring and supporting employee mental health, recognizing stress indicators, and offering timely interventions to promote well-being. Overall, the future landscape suggests a continued integration of AI technologies, emphasizing individualized experiences, well-being, ethical considerations, and adaptability to evolving work structures (Vrontis et al., 2022).

15.3.2 AI AND EMPLOYEE SATISFACTION

The increased use of AI has significantly impacted employee job satisfaction (Bhargava et al., 2021). Different AI tools adopted by organizations have changed job satisfaction levels, both positively and negatively. The positive effect is how AI-powered applicant tracking systems (ATS) help HR professionals streamline recruitment by analyzing resumes, identifying top candidates, and automating initial screenings. It saves time, allowing HR personnel to focus on more strategic tasks like engaging with candidates and improving the overall candidate experience, leading to higher job satisfaction among HR professionals due to reduced administrative burdens. Many tech companies utilize AI-powered tools for recruitment, such as IBM's Watson Recruitment or HireVue's AI-based video interviewing, to streamline processes and identify top talent. Similarly, AI-driven predictive analytics tools analyze large volumes of employee data to predict turnover risks. HR teams proactively take steps to retain valuable employees, address potential issues, and enhance job satisfaction using these tools. Also, AI-based learning platforms offer personalized training programs tailored to individual employee needs and learning styles. It enhances skill development and increases employee satisfaction by providing relevant and engaging learning experiences.

On the contrary, the negative effects of AI algorithms in recruitment inadvertently perpetuate biases in historical data, leading to discriminatory hiring practices. If historical data used to train algorithms reflect biases (e.g., gender or race), the AI system might unintentionally screen out qualified candidates based on those biased patterns (Ntoutsi et al., 2020). For example, if a tech company's historical hiring patterns favored male candidates, an AI system might learn this bias and continue prioritizing male applicants, thus perpetuating gender disparities and decreasing diversity in the workforce. The key to mitigating the adverse effects of AI in automated recruitment processes involves continuous monitoring and auditing of AI algorithms to ensure they do not reinforce biases, providing transparency in decision-making, and incorporating human oversight to intervene and rectify any biased outcomes (Ma and Sun, 2020).

Excessive reliance on AI in HR decision-making processes without human oversight creates employee mistrust (Chhillar and Aguilera, 2022). For instance, if an AI-driven system calculates bonuses without accounting for qualitative aspects like teamwork or innovation, employees might feel their efforts are undervalued, leading to frustration and demotivation. Similarly, if AI systems are solely relied upon for patient allocation to healthcare professionals without considering individual strengths or preferences, it can lead to dissatisfaction among healthcare workers. The fear of AI and automation replacing human jobs in HR functions creates anxiety and job insecurity among employees (Bhargava et al., 2021; McClure, 2018). Manufacturing firms in the automotive sector have experienced significant automation, where AI-driven robotics and machines have taken over assembly line tasks that humans previously did. This automation has led to job displacement for some workers, causing fear among employees in similar roles that their jobs might be at risk. This fear of job loss due to AI and automation creates dissatisfaction among workers (Dekker et al., 2017).

15.3.3 STRATEGIES FOR MAINTAINING POSITIVE WORKPLACE CULTURE IN THE AGE OF AI

The fear of AI replacing human jobs can have a profound psychological impact on employees, affecting their motivation, job engagement, and overall satisfaction. To mitigate these concerns, organizations must provide reskilling and upskilling opportunities, create a supportive environment for employees to adapt to technological changes, and emphasize the value of human skills that complement AI, such as creativity, emotional intelligence, and problem-solving. Proactively addressing these concerns helps alleviate fears and improve job satisfaction among employees affected by AI-driven changes in their industries (McClure, 2018).

1. **Emphasize Continuous Learning and Upskilling:** Encourage ongoing learning and upskilling programs to help employees adapt to technological changes (Jaiswal et al., 2022; Sofia et al., 2023). Provide resources for developing new skills that complement AI, such as creativity, emotional intelligence, and problem-solving. For example, in the tech industry, where rapid advancements occur, implementing continuous learning programs ensures that employees stay updated with the latest technologies. It fosters a culture of innovation, increases employee engagement, and enables the workforce to contribute meaningfully to technological advancements.

Microsoft launched the AI Academy to upskill its workforce in AI-related fields. This initiative provided various learning paths, including AI ethics and development, enabling employees to adapt to the AI-driven future. The program helped empower employees to leverage AI effectively while fostering a culture of continuous learning and innovation within the organization. Similarly, Tata Consultancy Services (TCS) initiated a large-scale upskilling program to prepare its workforce for AI and other emerging technologies. The program focused on retraining employees in AI-related fields, ensuring they were equipped with the necessary skills for the evolving technological landscape.

2. **Promote Collaboration between Humans and AI:** Encourage collaboration between AI systems and employees to optimize outcomes (Parker and Grote, 2022). Emphasize the complementary nature of AI and human capabilities, highlighting how they can work together effectively (Jarrahi, 2018). For example, AI tools can assist medical professionals in diagnoses and treatment recommendations in healthcare. Emphasizing collaboration between AI systems and healthcare workers enhances accuracy in patient care, reduces administrative burdens, and improves overall patient outcomes, leading to higher job satisfaction among healthcare professionals.

   IBM's Watson Health AI assists doctors by analyzing vast amounts of medical data to recommend treatment plans, aiding in more accurate diagnoses. This collaboration improved patient care outcomes while enabling healthcare professionals to make informed decisions, showcasing the power of AI-human collaboration in the healthcare industry. Similarly, Cellulant, a FinTech company in Kenya, has utilized AI-driven platforms to enhance financial services across Africa. Their focus on developing AI-based payment solutions tailored to the African market showcases how AI adoption drives financial inclusion and economic growth while fostering a positive workplace culture centered on innovation and inclusion.

3. **Transparent Communication and Decision-Making:** Maintain transparency in AI implementation, decision-making processes, and how AI impacts job roles. Ensure clear communication

about the rationale behind AI-driven decisions and its intended benefits (Yablonsky, 2021). For example, in the finance industry, where AI is used for risk assessment or investment decisions, transparent communication about AI algorithms and their role in decision-making helps build employee trust. This transparency fosters confidence in the technology and ensures that employees understand how AI complements their roles rather than replacing them, thus maintaining a positive workplace culture.

Google established AI principles to ensure transparency and responsible AI usage across its products. These principles guide decision-making processes, emphasizing transparency in AI development and deployment. By openly communicating about AI systems and their intended impacts, Google has aimed to build trust with its employees and users, emphasizing the importance of ethical and transparent AI implementations. InovaBH, a healthcare innovation hub in Brazil, has implemented AI-driven solutions in the healthcare sector. They have developed AI-powered tools for predictive analytics to optimize hospital operations and improve patient care. InovaBH's commitment to using AI to enhance healthcare services reflects a positive workplace culture focused on leveraging technology for societal benefit.

4. **Focus on Diversity and Mitigating Bias in AI:** Actively work to mitigate biases in AI algorithms and ensure diversity in its development teams to create fair and inclusive AI systems (Chhillar and Aguilera, 2022; Ntoutsi et al., 2020). For example, in retail, AI-driven recruitment tools are used extensively. Mitigating biases in these tools ensures fairer hiring practices, leading to a more diverse and inclusive workforce. It aligns with ethical standards and improves employee morale and satisfaction by fostering a more inclusive workplace culture. Accenture developed an AI fairness tool to reduce biases in AI algorithms used for recruitment. The tool helps identify and mitigate biases in hiring systems, promoting fairer and more inclusive hiring practices. This initiative demonstrates how organizations can actively work to create fair AI systems, contributing to a more diverse and inclusive workplace.

5. **Human-Centric Approach to Problem-Solving:** Encourage a problem-solving approach that values human input and creativity alongside AI capabilities (Jarrahi, 2018; Pandey et al., 2023). Emphasize the importance of human judgment in complex decision-making processes (Spector and Ma, 2019). For example, combining AI-driven predictive maintenance systems with human expertise in manufacturing allows for more effective problem-solving. It reduces downtime, improves productivity, and empowers employees to contribute their insights, enhancing job satisfaction and fostering a positive workplace culture. Siemens integrated AI-powered predictive maintenance systems in its manufacturing processes. It allowed for proactive maintenance by predicting machinery failures, combining AI's capabilities with human expertise for effective problem-solving. This approach increased operational efficiency, reduced downtime, and empowered employees to contribute their insights, showcasing the success of a human-centric approach to AI implementation in manufacturing.

15.4 IMPORTANCE OF THE HUMAN TOUCH IN HR

In the ever-evolving landscape of HRM, artificial intelligence (AI) has undeniably transformed how we approach talent acquisition, performance evaluation, and employee engagement (Allal-Chérif et al., 2021; Dwivedi et al., 2021). As organizations increasingly embrace these technological marvels for their efficiency and accuracy, a pivotal question arises: Amidst this surge of AI-powered innovations, where does the quintessential human touch stand in the realm of HR? The essence of human connection, empathy, and understanding remains an indispensable cornerstone in the HR domain, wielding an unparalleled capacity to foster trust, nurture relationships, and inspire a sense of belonging within the workforce (Budhwar et al., 2023; Vrontis et al., 2022). Here are several vital facets highlighting the indispensability of human interaction despite the increased use of AI:

1. **Emotional Intelligence and Empathy**: Humans possess a unique ability to understand, empathize, and connect with other individuals on an emotional level (Walker and Venker Weidenbenner, 2019). Emotional intelligence is crucial in HRM, especially in

conflict resolution, employee counselling, or support during challenging situations (Ashkanasy and Daus, 2002). AI lacks emotional capabilities and cannot truly empathize or comprehend complex human emotions, making the human touch indispensable in fostering trust and a supportive work environment (Beck and Libert, 2017; Prentice et al., 2020). For example, in the healthcare industry, HR professionals often deal with emotionally challenging situations involving employees, such as burnout, stress, or handling sensitive patient-related issues. While AI-driven systems can track workloads or offer generic wellness advice, human HR professionals provide empathetic support, actively listen to concerns, and tailor solutions that consider individual emotional needs, fostering a supportive environment for healthcare staff (Schuller and Schuller, 2018).

2. **Complex Decision-Making and Contextual Understanding**: AI systems excel in processing vast amounts of data and offering insights based on algorithms. However, they struggle with understanding nuanced situations, context, or cultural intricacies within an organization (Pillai et al., 2023; Sofia et al., 2023). Human interaction allows HR professionals to consider subjective factors, individual circumstances, and cultural nuances when making crucial decisions related to hiring, promotions, or addressing employee concerns (Jackson et al., 2014; Shen et al., 2009). When making decisions related to international assignments or cross-cultural training, AI systems might assist in providing data on market trends or language translation, but human HR experts play a crucial role in understanding the contextual complexities, cultural differences, and interpersonal dynamics, ensuring successful cross-border collaborations.

3. **Building Relationships and Employee Engagement**: Human interaction is fundamental in building strong relationships between employees and the organization. Personal interactions, mentoring, and coaching sessions establish rapport, trust, and loyalty, contributing significantly to employee engagement and retention (Mer, 2023). AI assists in providing information, but it lacks the personal touch required to cultivate meaningful connections (Chaudhuri and Ghosh, 2012). Despite relying on AI for

various functions, tech companies emphasize the human touch in fostering employee engagement. HR organizes regular mentorship programs, team-building activities, and one-on-one coaching sessions that strengthen employee interpersonal bonds (Mer, 2023). These personal interactions create a sense of belonging, enhancing retention and motivation among tech-savvy employees (Rubel et al., 2020).

4. **Effective Communication and Feedback**: Direct communication and personalized feedback are essential in nurturing employee growth and development. While AI facilitates feedback processes to some extent, it cannot often provide nuanced, constructive criticism or positive reinforcement tailored to an individual's unique strengths and weaknesses (Beck and Libert, 2017; Walker and Venker Weidenbenner, 2019). For example, in the retail sector, where turnover rates can be high, HR professionals supplement AI-driven performance reviews with personalized feedback sessions. Human interaction allows for constructive discussions addressing individual employee progress, areas of improvement, and career aspirations. This personalized approach contributes significantly to skill development and employee satisfaction, ultimately impacting customer service quality (Sofia et al., 2023).
5. **Ethical and Moral Decision-Making**: Human judgment is crucial in navigating ethical dilemmas or making moral decisions within the workplace (Nassar and Kamal, 2021; Sekerka et al., 2009). AI operates based on pre-defined algorithms without considering ethical implications or moral aspects of decision-making. Financial institutions often face ethical dilemmas regarding compliance, data privacy, and customer trust. While AI systems aid in risk assessment and compliance checks, human HR professionals are critical in interpreting regulations, navigating gray areas, making decisions aligned with ethical standards, and preserving institutional integrity.
6. **Adaptability and Creativity**: HRM often requires adaptability and creativity, especially in designing strategies for talent development, organizational culture enhancement, or addressing unforeseen challenges. Human interaction brings adaptability and creative problem-solving skills that AI, driven by programmed

algorithms, struggles to emulate (Colin et al., 2016). For example, in the entertainment industry, HR professionals work on talent development and retention strategies. While AI algorithms might assist in predicting market trends or audience preferences, human creativity plays a vital role in devising innovative strategies for content creation, talent management, employee creativity, and fostering a dynamic work culture that resonates with creative individuals.

15.5 AREAS WHERE HUMAN INVOLVEMENT REMAINS CRITICAL

While AI and technological advancements continue to reshape various facets of HRM, several core HR functions inherently require human involvement due to their nuanced, interpersonal, and strategic nature. Here are key HR functions where human involvement remains critical:

1. **Talent Acquisition and Recruitment**: Human involvement is crucial in the initial stages of talent acquisition. While AI helps in sourcing candidates, analyzing resumes, or conducting initial screenings based on specific criteria, human HR professionals bring contextual understanding, emotional intelligence, and intuition to assess cultural fit, soft skills, and potential beyond what is listed on a resume (Allal-Chérif et al., 2021; Rehman et al., 2022). They build relationships with candidates, conduct interviews, and make subjective judgments critical for finding the right talent aligned with the organization's values and goals.
2. **Employee Development and Training**: While AI facilitates personalized learning paths and recommends training modules based on data analysis, human trainers or facilitators play a pivotal role in delivering engaging, interactive learning experiences (Bhutoria, 2022; Chen, 2023). They create a conducive environment for skill development, offer real-time feedback, and personalize training to address individual learning styles or areas needing improvement. Human trainers adapt content on the fly and provide context that resonates with employees, enhancing the effectiveness of learning initiatives (Hollweck and Lofthouse, 2021).

3. **Employee Engagement and Relationship Management**: Maintaining a motivated and engaged workforce requires human interaction and connection. HR professionals foster engagement through regular communication, team-building activities, and recognition programs (Chaudhuri and Ghosh, 2012). They listen to employee concerns, mediate conflicts, and offer support, building trust and loyalty within the workforce (Abbass, 2019). Human HR practitioners understand employees' unique needs and tailor engagement strategies accordingly, ensuring a positive work environment.
4. **Performance Management and Feedback**: While AI systems track performance metrics and offer data-driven insights, human involvement is indispensable in providing meaningful feedback. HR professionals conduct performance reviews, set goals collaboratively, and offer personalized feedback that considers individual strengths, areas for improvement, and career aspirations (Budhwar et al., 2023; Shen et al., 2009). Human interaction fosters a supportive environment for ongoing dialogue, coaching, and development planning (Hollweck and Lofthouse, 2021).
5. **Organizational Culture and Change Management**: Shaping and maintaining organizational culture requires human leadership and interaction (Marion and Fixson, 2021). HR professionals drive cultural initiatives, facilitate change management, and embody cultural values through actions and interactions. They serve as cultural ambassadors, guiding employees through transitions, fostering inclusivity, and ensuring that company culture aligns with organizational goals (Akinnusi et al., 2017).
6. **Employee Well-Being and Conflict Resolution**: Handling employee concerns, conflicts, and well-being matters necessitates human empathy and understanding (Prentice et al., 2020; Walker and Venker Weidenbenner, 2019). HR professionals provide a safe space for employees to voice concerns, offer support during challenging times, and mediate conflicts impartially. Their ability to listen, empathize, and navigate complex human emotions contributes significantly to employee satisfaction and retention.

These core HR functions heavily rely on human involvement due to their interpersonal, subjective, and strategic nature (Budhwar et al., 2023;

Strohmeier and Piazza, 2015; Vrontis et al., 2022). While AI complements these functions by providing data-driven insights and streamlining processes, the human touch remains indispensable in fostering relationships, understanding human emotions, and making subjective decisions critical for effective people management within organizations (Abbass, 2019; Jaiswal et al., 2022; Pandey et al., 2023; Yablonsky, 2021).

15.6 FUTURE OF AI AND HR COLLABORATION

As we stand at the precipice of a new era in HRM, the convergence of AI and human expertise heralds a transformative partnership that holds immense promise for the future of the workplace. We must understand the symbiotic relationship between AI technology and human ingenuity within HRM, illuminating the burgeoning landscape where automation meets empathy, data-driven insights converge with intuitive decision-making, and innovation merges with the human touch (Bhutoria, 2022; Spector and Ma, 2019). As organizations navigate the evolving dynamics of the digital age, this exploration aims to unveil the unprecedented potentials and collaborative possibilities that lie ahead in shaping a progressive and harmonious alliance between AI and human-centric HR practices (Chowdhury et al., 2023; Mer, 2023; Rodgers et al., 2023).

15.6.1 STRATEGIES FOR PREPARING HR PROFESSIONALS FOR FUTURE AI DEVELOPMENTS

Preparing HR professionals for the future AI-driven landscape involves strategic planning, upskilling, and fostering a mindset that embraces technological advancements while retaining human-centric values. Here are comprehensive strategies for preparing HR professionals for AI developments:

1. **Education and Training Initiatives**: Implement structured training programs and workshops on AI fundamentals, data literacy, and understanding AI's impact on HRM (Chen, 2023). These programs should cover AI ethics, bias mitigation, and how AI augments rather than replaces human roles. Encourage HR professionals to participate in online courses, certifications, or seminars to stay updated on AI advancements.

2. **Developing Digital and Analytical Skills**: Equip HR teams with digital fluency and analytical skills essential for leveraging AI-driven tools (Böhmer and Schinnenburg, 2023). Encourage proficiency in data analytics, interpretation of AI-generated insights, and the ability to derive actionable strategies from AI-generated data (Nassar and Kamal, 2021). Provide hands-on experiences with AI-powered HR platforms and tools to build confidence and competence.
3. **Cultivating Adaptability and Change Management Skills**: Foster adaptability among HR professionals to embrace technological changes (Rubel et al., 2020). Develop change management capabilities to facilitate smooth transitions when integrating AI into HR processes. Encourage a culture that values innovation, experimentation, and continuous learning to effectively adapt to evolving AI technologies.
4. **Emphasizing Soft Skills and Emotional Intelligence**: Highlight the significance of human-centric skills such as empathy, emotional intelligence, and interpersonal communication (Prentice et al., 2020). AI handles routine tasks, but human HR professionals excel in understanding emotions, resolving conflicts, and fostering relationships. Encourage the development of these skills to complement AI's technical capabilities.
5. **Encouraging Collaboration between HR and Technology Teams:** Facilitate collaboration between HR and technology departments to ensure HR professionals understand AI capabilities and limitations (Chowdhury et al., 2023). Create cross-functional teams to jointly explore, evaluate, and implement AI solutions tailored for HR functions. This collaboration ensures alignment between AI technology and HR goals.
6. **Promoting Ethical Use of AI in HRM**: Instill a solid ethical foundation emphasizing fairness, transparency, and accountability in AI implementation (Nassar and Kamal, 2021; Rodgers et al., 2023). HR professionals should understand the ethical implications of AI, including bias mitigation, privacy concerns, and ethical decision-making in HR processes. Develop guidelines and protocols for ethical AI use within HR practices.

7. **Continuous Monitoring and Evaluation**: Establish mechanisms for continuous evaluation and feedback loops to assess the impact of AI implementations in HRM (Vrontis et al., 2022). Regularly gather insights from HR professionals about AI's effectiveness, identify areas for improvement, and iterate strategies accordingly. Encourage open dialogue and a feedback culture to refine AI integration strategies (Abbass, 2019).
8. **Leadership Support and Advocacy**: Secure leadership support to champion AI initiatives within HR (Chowdhury et al., 2023). Leaders should advocate for AI adoption, emphasize its value, and encourage HR professionals to embrace AI as a tool that complements their expertise rather than a replacement (Pillai and Sivathanu, 2020). Leadership endorsement creates an environment conducive to AI-driven HR innovation.

15.7 CONCLUSION

This chapter explains the transformative landscape of HRM in the age of AI. It explains the evolution of HR practices, exploring how AI has streamlined processes, offered data-driven insights, and reshaped traditional functions. However, amidst the technological leaps, a resounding theme persists—the irreplaceable value of the human element in HRM. AI has undoubtedly revolutionized talent acquisition, learning and development, employee engagement, and decision-making processes within HR. Nevertheless, it is essential to recognize that while AI enhances efficiency, it lacks the innate qualities that define the human touch: empathy, emotional intelligence, adaptability, and ethical reasoning. The fusion of AI with the human element is not about pitting one against the other but rather about striking a harmonious balance, where technology amplifies human capabilities rather than replaces them. HR professionals must be equipped with the necessary skills—digital literacy, adaptability, ethical awareness, and interpersonal skills—to leverage AI effectively. They are not just operators of AI systems but orchestrators, ensuring that technology aligns with organizational values, fosters inclusivity, and enhances the employee experience. As organizations march into an AI-driven future, the ongoing challenge lies in maintaining equilibrium—a delicate balance where AI optimizes processes, provides insights, and augments decision-making.

At the same time, the human touch remains the cornerstone of empathy, connection, and ethical judgment. Achieving this balance will pave the way for a future where HR practices are efficient, empathetic, inclusive, and geared towards nurturing the best in individuals and organizations.

KEYWORDS

- **digital wealth management**
- **FinTech transformation**
- **portfolio optimization**
- **robo-advisors**
- **sustainable finance**

REFERENCES

Abbass, H. A. (2019). Social integration of artificial intelligence: Functions, automation allocation logic and human-autonomy trust. *Cognitive Computation, 11*(2), 159–171.

Akinnusi, D. M., Sonubi, O. O., & Oyewunmi, A. E. (2017). Fostering effective workforce diversity management in Nigerian organizations: The challenge of human resource management. *International Review of Management and Marketing, 7*(2), 108–116.

Allal-Chérif, O., Aranega, A. Y., & Sánchez, R. C. (2021). Intelligent recruitment: How to identify, select, and retain talents from around the world using artificial intelligence. *Technological Forecasting and Social Change, 169*, 120822.

Ashkanasy, N. M., & Daus, C. S. (2002). Emotion in the workplace: The new challenge for managers. *Academy of Management Perspectives, 16*(1), 76–86.

Beck, M., & Libert, B. (2017). The rise of AI makes emotional intelligence more important. *Harvard Business Review, 15*(1–5).

Bhargava, A., Bester, M., & Bolton, L. (2021). Employees' perceptions of the implementation of robotics, artificial intelligence, and automation (RAIA) on job satisfaction, job security, and employability. *Journal of Technology in Behavioral Science, 6*(1), 106–113.

Bhutoria, A. (2022). Personalized education and artificial intelligence in the United States, China, and India: A systematic review using a human-in-the-loop model. *Computers and Education: Artificial Intelligence, 3*, 100068.

Böhmer, N., & Schinnenburg, H. (2023). Critical exploration of AI-driven HRM to build up organizational capabilities. *Employee Relations: The International Journal.*

Budhwar, P., Chowdhury, S., Wood, G., Aguinis, H., Bamber, G. J., Beltran, J. R., Boselie, P., Lee Cooke, F., Decker, S., & DeNisi, A. (2023). Human resource management in the age of generative artificial intelligence: Perspectives and research directions on ChatGPT. *Human Resource Management Journal, 33*(3), 606–659.

Chaudhuri, S., & Ghosh, R. (2012). Reverse mentoring: A social exchange tool for keeping the boomers engaged and millennials committed. *Human Resource Development Review, 11*(1), 55–76.

Chen, Z. (2023). Artificial intelligence-virtual trainer: Innovative didactics aimed at personalized training needs. *Journal of the Knowledge Economy, 14*(2), 2007–2025.

Chhillar, D., & Aguilera, R. V. (2022). An eye for artificial intelligence: Insights into the governance of artificial intelligence and vision for future research. *Business & Society, 61*(5), 1197–1241.

Chowdhury, S., Dey, P., Joel-Edgar, S., Bhattacharya, S., Rodriguez-Espindola, O., Abadie, A., & Truong, L. (2023). Unlocking the value of artificial intelligence in human resource management through AI capability framework. *Human Resource Management Review, 33*(1), 100899.

Colin, T. R., Belpaeme, T., Cangelosi, A., & Hemion, N. (2016). Hierarchical reinforcement learning as creative problem solving. *Robotics and Autonomous Systems, 86*, 196–206.

Dekker, F., Salomons, A., & van der Waal, J. (2017). Fear of robots at work: The role of economic self-interest. *Socio-Economic Review, 15*(3), 539–562.

Dwivedi, Y. K., Hughes, L., Ismagilova, E., Aarts, G., Coombs, C., Crick, T., Duan, Y., Dwivedi, R., Edwards, J., & Eirug, A. (2021). Artificial intelligence (AI): Multidisciplinary perspectives on emerging challenges, opportunities, and agenda for research, practice and policy. *International Journal of Information Management, 57*, 101994.

Garg, S., Sinha, S., Kar, A. K., & Mani, M. (2022). A review of machine learning applications in human resource management. *International Journal of Productivity and Performance Management, 71*(5), 1590–1610.

Hollweck, T., & Lofthouse, R. M. (2021). Contextual coaching: Levering and leading school improvement through collaborative professionalism. *International Journal of Mentoring and Coaching in Education, 10*(4), 399–417.

Jackson, S. E., Schuler, R. S., & Jiang, K. (2014). An aspirational framework for strategic human resource management. *Academy of Management Annals, 8*(1), 1–56.

Jaiswal, A., Arun, C. J., & Varma, A. (2022). Rebooting employees: Upskilling for artificial intelligence in multinational corporations. *The International Journal of Human Resource Management, 33*(6), 1179–1208.

Jarrahi, M. H. (2018). Artificial intelligence and the future of work: Human-AI symbiosis in organizational decision making. *Business Horizons, 61*(4), 577–586.

Ma, L., & Sun, B. (2020). Machine learning and AI in marketing: Connecting computing power to human insights. *International Journal of Research in Marketing, 37*(3), 481–504.

Marion, T. J., & Fixson, S. K. (2021). The transformation of the innovation process: How digital tools are changing work, collaboration, and organizations in new product development. *Journal of Product Innovation Management, 38*(1), 192–215.

McClure, P. K. (2018). "You're fired," says the robot: The rise of automation in the workplace, technophobes, and fears of unemployment. *Social Science Computer Review, 36*(2), 139–156.

Mer, A. (2023). Artificial intelligence in human resource management: Recent trends and research agenda. *Digital Transformation, Strategic Resilience, Cyber Security and Risk Management, 111*, 31–56.

Munoko, I., Brown-Liburd, H. L., & Vasarhelyi, M. (2020). The ethical implications of using artificial intelligence in auditing. *Journal of Business Ethics, 167*, 209–234.

Nassar, A., & Kamal, M. (2021). Ethical dilemmas in AI-powered decision-making: A deep dive into big data-driven ethical considerations. *International Journal of Responsible Artificial Intelligence, 11*(8), 1–11.

Ntoutsi, E., Fafalios, P., Gadiraju, U., Iosifidis, V., Nejdl, W., Vidal, M. E., Ruggieri, S., Turini, F., Papadopoulos, S., & Krasanakis, E. (2020). Bias in data-driven artificial intelligence systems—An introductory survey. *Wiley Interdisciplinary Reviews: Data Mining and Knowledge Discovery, 10*(3), e1356.

Pandey, A., Panday, S. P., & Joshi, B. (2023). Design and development of applications using human-computer interaction. In *Innovations in Artificial Intelligence and Human-Computer Interaction in The Digital Era* (pp. 255–293). Elsevier.

Parker, S. K., & Grote, G. (2022). Automation, algorithms, and beyond: Why work design matters more than ever in a digital world. *Applied Psychology, 71*(4), 1171–1204.

Pianese, T., Errichiello, L., & da Cunha, J. V. (2023). Organizational control in the context of remote working: A synthesis of empirical findings and a research agenda. *European Management Review, 20*(2), 326–345.

Pillai, R., & Sivathanu, B. (2020). Adoption of artificial intelligence (AI) for talent acquisition in IT/ITeS organizations. *Benchmarking: An International Journal, 27*(9), 2599–2629.

Pillai, R., Ghanghorkar, Y., Sivathanu, B., Algharabat, R., & Rana, N. P. (2023). Adoption of artificial intelligence (AI) based employee experience (EEX) chatbots. *Information Technology & People.*

Prentice, C., Dominique Lopes, S., & Wang, X. (2020). Emotional intelligence or artificial intelligence–An employee perspective. *Journal of Hospitality Marketing & Management, 29*(4), 377–403.

Rehman, S., Ullah, A., Naseem, K., Elahi, A. R., & Erum, H. (2022). Talent acquisition and technology: A step towards sustainable development. *Frontiers in Psychology, 13*, 979991.

Rodgers, W., Murray, J. M., Stefanidis, A., Degbey, W. Y., & Tarba, S. Y. (2023). An artificial intelligence algorithmic approach to ethical decision-making in human resource management processes. *Human Resource Management Review, 33*(1), 100925.

Rubel, M. R. B., Kee, D. M. H., & Rimi, N. N. (2020). Matching people with technology: Effect of HIWP on technology adaptation. *South Asian Journal of Human Resources Management, 7*(1), 9–33.

Schuller, D., & Schuller, B. W. (2018). The age of artificial emotional intelligence. *Computer, 51*(9), 38–46.

Sekerka, L. E., Bagozzi, R. P., & Charnigo, R. (2009). Facing ethical challenges in the workplace: Conceptualizing and measuring professional moral courage. *Journal of Business Ethics, 89*, 565–579.

Shen, J., Chanda, A., D'Netto, B., & Monga, M. (2009). Managing diversity through human resource management: An international perspective and conceptual framework. *The International Journal of Human Resource Management, 20*(2), 235–251.

Sofia, M., Fraboni, F., De Angelis, M., Puzzo, G., Giusino, D., & Pietrantoni, L. (2023). The impact of artificial intelligence on workers' skills: Upskilling and reskilling in organizations. *Informing Science: The International Journal of an Emerging Trans discipline, 26*, 39–68.

Spector, J. M., & Ma, S. (2019). Inquiry and critical thinking skills for the next generation: From artificial intelligence back to human intelligence. *Smart Learning Environments, 6*(1), 1–11.

Strohmeier, S., & Piazza, F. (2015). Artificial intelligence techniques in human resource management—A conceptual exploration. In *Intelligent Techniques in Engineering Management: Theory and Applications* (pp. 149–172).

Vrontis, D., Christofi, M., Pereira, V., Tarba, S., Makrides, A., & Trichina, E. (2022). Artificial intelligence, robotics, advanced technologies and human resource management: A systematic review. *The International Journal of Human Resource Management, 33*(6), 1237–1266.

Walker, G., & Venker Weidenbenner, J. (2019). Social and emotional learning in the age of virtual play: Technology, empathy, and learning. *Journal of Research in Innovative Teaching & Learning, 12*(2), 116–132.

Yablonsky, S. (2021). AI-driven platform enterprise maturity: From human led to machine governed. *Kybernetes, 50*(10), 2753–2789.

CHAPTER 16

Enhancing Robo-Advisor User Experience Through Financial Literacy and Apparent Trust

MEDHA GUPTA,[1,*] DIXIT KALRA,[2] and SWATI GUPTA[3]

[1]*University School of Business, Chandigarh University, Gharuan, Mohali, Punjab, India*

[2]*School of Management Studies, CGC Mohali, Punjab, India*

[3]*PML SD Business School, Chandigarh, India*

*Corresponding author

ABSTRACT

This chapter interrogates the determinants of robo-advisor adoption, with a particular focus on the interrelationship of financial literacy, apparent trust, and user-friendliness in influencing investor engagement. Employing a survey of 325 respondents analyzed through Partial Least Squares Structural Equation Modeling (PLS-SEM), the study reveals that financial literacy and trust are significant predictors of adoption, while user-friendliness exerts limited or even adverse influence. The findings illuminate that beyond intuitive interface design, the perceived credibility, transparency, and data security of robo-advisors constitute the core determinants of acceptance. Financial literacy emerges as a catalyst, empowering users to critically engage with algorithmic recommendations and enhancing their confidence in technology-mediated financial advice. Similarly, apparent trust—anchored in reliability and ethical governance—plays a decisive role in overcoming skepticism surrounding algorithmic

Robo-Advisors and Artificial Intelligence in Human Resources Management: Revolutionizing HR.
Sanjay Taneja, Swati Gupta, Girish Lakhera, Mohit Kukreti, and Ercan Özen (Eds.)

DOI: 10.1201/9781779641410-16

decision-making. The research contributes to both theoretical advancement and practical application by emphasizing that fostering digital trust ecosystems and enhancing financial education are indispensable for the sustainable adoption of robo-advisory platforms. Developers, regulators, and policymakers are urged to prioritize data security, ethical oversight, and literacy initiatives to ensure inclusive and responsible digital finance adoption across socio-economic strata.

16.1 INTRODUCTION

Robo-advisors are online tools that offer automated financial planning and investment management with minimal human involvement. They collect information on clients' financial situation and financial aspirations through a questionnaire online. From this information, they give a unique recommendation, as well as the investment of the client's funds, which, where applicable, happens automatically (Frankenfield, 2020). It is a revolutionary service that nurtures social interaction as well as strong economic sustainment (Cheng et al., 2019). It is possible to note that this creative service is beneficial for forming better conditions for societal welfare and sustainable economic development. In particular, it improves and optimizes the existing Robo-advisor products, using advanced methods based on big data, customer profiling, and the cloud. These improvements not only advance these technologies but also foster their broader application across various industries, driving further technological breakthroughs (Gupta et al., 2024). Additionally, by linking financial management and markets with sustainability (Kalra et al., 2024; Singla et al., 2024), the promotion of Robo-advisors supports the growth of the financial sector, boosts social dynamism, and fosters sustainable economic development (Castro and Chousa, 2006). However, due to the lack of standardized criteria in the industry, it is challenging for customers to trust Robo-advisor products, resulting in a slow rate of adoption (Mallat, 2007).

When consumers are involved, they tend to seek advice from human financial consultants. This may be because people feel calmer being advised by a person rather than a robot or artificial intelligence (AI), like in Robo-advisors (Northey et al., 2022). On the contrary, the Swedes, Norwegians, and Finns who use Robo advisors, and residents from less affluent audiences, women, and older people indicate their readiness to participate in sustainable investing. They also display low risk-taking propensity and have a short investment time frame. Nevertheless, the study excluded other possible related variables like financial literacy, digital literacy, and investment experience from the researchers' discussions (Faradynawati and Söderberg, 2022).

The study focuses on comparing the level of financial literacy, user-friendliness, and apparent trust between the two groups of participants to investigate how these factors affect the adoption and usage of Robo-advisors and gain valuable information about readiness to use them. The following research questions are posed: (1) How do the levels of familiarity with financial Robo-advisors affect the adoption intentions?; and (2) How much do user-friendliness and trustworthiness influence the likelihood of people adopting Robo-advisors?

16.2 LITERATURE REVIEW

16.2.1 APPARENT TRUST (AT)

The matter is that trust is crucial in the case of Robo-advisors and other AI services because these solutions are highly autonomous (Ostrom et al., 2019). In this vein, the elements of establishing social interaction and cultivating bonds with clients in the context of SST are reported to be missing (Bock et al., 2020). However, where new technology seems to be both useless and highly threatening to the original system, it will not be adopted unless new specific efforts are made to change the user's perception and attitude towards the new technology (Milani, 2019). New technologies must be safeguarded to prevent unauthorized access to financial data during the processes of transmission and storage; this type of data is very personal and sensitive (Tun-Pin et al., 2019). From this, we get our H1, which is as follows:

- **H1:** More specifically, it is ascertained that there is a presence of strong correlation between the level of apparent trust and the usage of Robo-advisors.

16.2.2 FINANCIAL LITERACY (FL)

It implies having the competence and proficiency, as well as the pattern of behaviors that enable individuals to make rational decisions within the areas of banking, borrowing, and investing, and the regular and efficient management of their day-to-day financial decisions, precisely and competently. Robo-advisors were designed to provide services at lower costs to generate interest from millennials who do not have much financial knowledge (Lusardi and Mitchell, 2014). In fact, studies show that mastery of financial literacy bears a huge influence on the kinds of decisions and actions made concerning money. For example, people who have good knowledge of financial activities are inclined towards engaging themselves in market activities and investing in stocks (Yi et al., 2023). From this, we get our H2, which is as follows:

- **H2:** In particular, there is quite a positive and meaningful relationship between financial literacy and the usage of Robo-advisors.

16.2.3 USER FRIENDLY (UF)

The expansion of Robo-advisory services will accelerate if prospective users find the system beneficial and user-friendly (Cendrell and Issa, 2018). People are more likely to embrace new technologies when they strongly believe these technologies will benefit them in some way (Fallon and Scherer, 2017). One notable aspect of Robo-advisors is their ability to offer a higher degree of customization compared to traditional human financial advisors (Gupta et al., 2023). The previous study also highlighted that adopting Robo-advisory services enhances customers' comprehension, which in turn expedites their power to make informed decisions about personal finance and investments (Belanche et al., 2019). From this, we get our H3, which is as follows:

- **H3:** There is an important association between user-friendliness and the robot advisors' adoption.

16.3 RESEARCH METHODOLOGY

Based on the review of the literature, we formulated the conceptual framework illustrated in Figure 16.1. Our study conducted an online survey with a sample of 325 investors, collecting all the questionnaires through the Internet. All items of measurement utilized a seven-point Likert scale spanning from Category 1 (Strongly Disagree) to Category 7 (Strongly Agree).

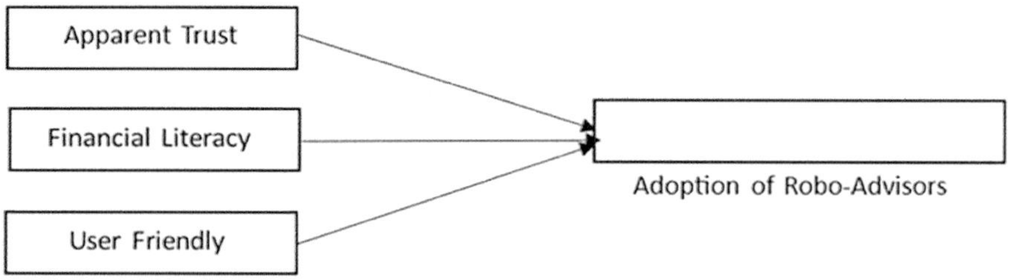

FIGURE 16.1 Conceptual framework.

Source: Authors Compilation.

16.4 RESULTS

We have used Smart PLS-SEM for the data analysis. Before formally testing our hypotheses, we evaluated the validity and reliability of our measurements through a three-step process. Initially, we examined the scales for our constructs using cross-loading analysis. Each factor demonstrated a loading greater than 0.7, indicating that each item aligned well with its corresponding latent construct. Subsequently, based on recommendations by Hair et al. (2011), we calculated Cronbach's alpha (CA), composite reliability (CR), and average variance extracted (AVE) as indicated in Table 16.1. Last but not least, to affirm discriminant validity, using the procedure by Wiersma (1994), each construct's square root of the AVE (SQAVE) was computed and compared with the absolute value of the corresponding construct's latent variable correlations. This means that there were no dominant constructs, as exemplified by the diagonal values in Table 16.2, which were greater than the values below them. Based on this discussion, we established that the developed questionnaire is both reliable and valid.

To assess the proposed research hypotheses, the chapter uses the method of partial least squares structural equation modeling (PLS-SEM) (*See* Figure 16.2). After completing the reliability and validity of the scale, we employed the standard bootstrapping procedure to derive path coefficients. To substantiate the presumptions postulated for the research and establish the relevance and significance of the proposed relationships, we computed mean (M), standard deviation (STDEV), t-value, and p-value. The tests' results are summarized in Table 16.3.

TABLE 16.1 Construct Reliability and Validity Test

Construct	Statements	Factor Loadings	Cronbach's Alpha	Composite Reliability	Average Variance Extracted
Apparent Trust (AT)	AP1	0.946	0.950	0.964	0.869
	AP2	0.925			
	AP3	0.953			
	AP4	0.904			
Financial Literacy (FL)	FL1	0.789	0.864	0.908	0.711
	FL2	0.904			
	FL3	0.863			
	FL4	0.812			
User Friendly (UF)	UF1	0.926	0.953	0.966	0.875
	UF2	0.904			
	UF3	0.950			
	UF4	0.962			
Robo Advisor (RB)	RB1	0.903	0.960	0.968	0.836
	RB2	0.839			
	RB3	0.962			
	RB4	0.954			
	RB5	0.910			
	RB6	0.912			

Source: Author's Compilation.

TABLE 16.2 Discriminant Validity Test

		Heterotrait–Monotrait Ratio (HTMT)		
	AT	**FL**	**RB**	**IF**
AT				
FL	0.424			
RB	0.528	0.668		
IF	0.356	0.787	0.309	
		Fornell–Larcker Criterion		
	AT	**FL**	**RB**	**IF**
AT	0.932			
FL	0.405	0.843		
RB	0.514	0.615	0.914	
IF	0.331	0.712	0.317	0.936

Source: Author's Compilation.

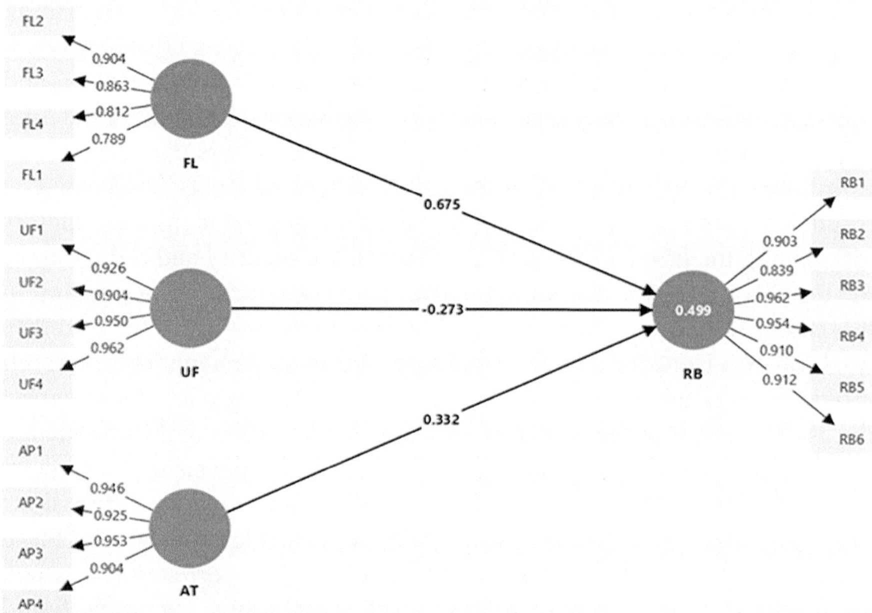

FIGURE 16.2 PLS-SEM model.

Source: Author's compilation.

TABLE 16.3 Hypothesis Testing Results

Hypothesis	Path	Sample Mean	Standard Deviation	T Statistics	P Value	Result
H_1	AT✍RB	0.370	0.093	4.017	0.000	Accepted
H_2	FL✍RB	0.680	0.170	3.962	0.000	Accepted
H_3	UF✍RB	–0.203	0.220	1.240	0.215	Rejected

Source: Author's Compilation.

Based on the data in Table 16.3, we can conclude that hypotheses 1 and 2 were confirmed, demonstrating a positive relationship between both apparent trust and financial literacy with the adoption of Robo-advisors. In contrast, hypothesis 3 was not supported, suggesting that user-friendliness negatively impacts the adoption of Robo-advisors.

16.5 CONCLUSION

Robo-advisors have become strong competitors in the competitive and ever-changing field of financial advice because they solve the problem of individualization of investment recommendations using an algorithm. The executed work analyzes the factors encouraging Robo-Advisor usage, with special emphasis on discretion, financial competence, and user-friendliness. Procedures such as structural equation modeling (SEM) will be employed to determine how these factors connect as well as influence users' intention to avail themselves of Robo-Advisors. The results confirm that both trust and financial literacy are significant factors influencing the usage of the service. Trust in the security and reliability of the Robo-advisors for operating the user's investments becomes central in this case. Furthermore, high financial literacy increases user employability of Robo-Advisors for efficient working and management. The research provides practical applications for developers and marketers by stressing the establishment of the confidence of users and enhancing the knowledge of potential users on financial aspects. Besides, it suggests that designers should pay close attention to interfaces as they can influence impressions of the organization as competent and reliable. Possible extensions of these studies may be associated with the examination of the correlation between the level of user-friendliness and the level of adoptions: it may be found that there are different groups of users with different financial perceptions. Such an understanding could help design specific ways of improving the uptake of such services across the demographic categories of consumers.

KEYWORDS

- **apparent trust**
- **digital trust ecosystem**
- **financial literacy**
- **robo-advisors**
- **user adoption**

REFERENCES

Belanche, D., Casaló, L. V., & Flavián, C. (2019). Artificial intelligence in FinTech: Understanding robo-advisors adoption among customers. *Industrial Management & Data Systems, 119*(7), 1411–1430.

Bock, D. E., Wolter, J. S., & Ferrell, O. C. (2020). Artificial intelligence: Disrupting what we know about services. *Journal of Services Marketing, 34*(3), 317–334.

Castro, N. R., & Chousa, J. P. (2006). An integrated framework for the financial analysis of sustainability. *Business Strategy and the Environment, 15*(5), 322–333.

Cedrell, L., & Issa, N. (2018). The adoption of robo-advisory in the Swedish financial technology market: Analyzing the consumer perspective.

Cheng, X., Guo, F., Chen, J., Li, K., Zhang, Y., & Gao, P. (2019). Exploring the trust influencing mechanism of robo-advisor service: A mixed method approach. *Sustainability, 11*(18), 4917. https://doi.org/10.3390/su11184917 (accessed on 11th August 2025).

Faloon, M., & Scherer, B. (2017). Individualization of robo-advice. *The Journal of Wealth Management, 20*(1), 30. https://doi.org/10.3905/jwm.2017.20.1.030 (accessed on 11th August 2025).

Faradynawati, I. A. A., & Söderberg, I. L. (2022). Sustainable investment preferences among robo-advisor clients. *Sustainability, 14*(19), 12636.

Frankenfield, J. (2020). Cloud computing. *Investopedia*. https://www.investopedia.com/terms/c/cloud-computing.asp (accessed on 11th August 2025).

Gupta, M., Arora, R., & Taneja, S. (2023). Bibliometric analysis on employee engagement and human resource management. In *Enhancing Customer Engagement Through Location-Based Marketing* (pp. 185–205). IGI Global. https://doi.org/10.4018/978-1-6684-8177-6.ch013 (accessed on 11th August 2025).

Gupta, S., Kaur, G., Kumari, I., & Kalra, D. (2024). Exploration of opportunities and challenges in green finance: A journey of sustainable development. In *Sustainable Investments in Green Finance* (pp. 256–270). https://doi.org/10.4018/979-8-3693-1388-6.ch018 (accessed on 11th August 2025).

Kalra, D., Gupta, M., Gupta, S., Singh, H., & Agrawal, P. (2023). Exploring the impact of green marketing on consumer behavior in the FMCG industry: A sustainable path to consumer engagement. In *Green Management–A New Paradigm in the World of Business* (pp. 81–93).

Kalra, D., Gupta, M., Gupta, S., Singh, H., & Agrawal, P. (2024). Quantifying influence: A bibliometric analysis of robo-advisors in management. In *Robo-Advisors in Management* (pp. 175–186). https://doi.org/10.4018/979-8-3693-2849-1.ch012 (accessed on 11th August 2025).

Lusardi, A., & Mitchell, O. S. (2014). The economic importance of financial literacy: Theory and evidence. *American Economic Journal: Journal of Economic Literature, 52*(1), 5–44. https://doi.org/10.1257/jel.52.1.5 (accessed on 11th August 2025).

Malhotra, N., & Aggarwal, R. (2022). A review-based study on transformational leadership and its prospects. *ECS Transactions, 107*(1), 9173.

Mallat, N. (2007). Exploring consumer adoption of mobile payments–A qualitative study. *The Journal of Strategic Information Systems, 16*(4), 413–432. https://doi.org/10.1016/j.jsis.2007.08.001 (accessed on 11th August 2025).

Milani, A. (2019). The role of risk and trust in the adoption of robo-advisory in Italy: An extension of the unified theory of acceptance and use of technology. PricewaterhouseCoopers Advisory SpA.

Northey, G., Hunter, V., Mulcahy, R., Choong, K., & Mehmet, M. (2022). Man vs machine: How artificial intelligence in banking influences consumer belief in financial advice.

International Journal of Bank Marketing, 40(6), 1182–1199. https://doi.org/10.1108/IJBM-09-2021-0439 (accessed on 11th August 2025).

Ostrom, A. L., Fotheringham, D., & Bitner, M. J. (2019). Customer acceptance of AI in service encounters: Understanding antecedents and consequences. In *Handbook of Service Science* (Vol. II, pp. 77–103). https://doi.org/10.1007/978-3-319-98512-1_5 (accessed on 11th August 2025).

Singla, B., Shalender, K., & Singh, N. (2024). *Omnichannel Approach to Co-Creating Customer Experiences Through Metaverse Platforms*. IGI Global.

Tun-Pin, C., Keng-Soon, W. C., Yen-San, Y., Pui-Yee, C., Hong-Leong, J. T., & Shwu-Shing, N. (2019). An adoption of fintech service in Malaysia. *South East Asia Journal of Contemporary Business, 18*(5), 134–147.

Yi, T. Z., Rom, N. A. M., Hassan, N. M., Samsurijan, M. S., & Ebekozien, A. (2023). The adoption of robo-advisory among millennials in the 21st century: Trust, usability and knowledge perception. *Sustainability, 15*(7), 6016. https://doi.org/10.3390/su15076016 (accessed on 11th August 2025).

Index

For Product Safety Concerns and Information please contact our EU representative GPSR@taylorandfrancis.com Taylor & Francis Verlag GmbH, Kaufingerstraße 24, 80331 München, Germany